Dr Norman Lazarus is a qualified medical doctor and scientist. He has worked in academia, in industry with the Wellcome Foundation and in the Department of Health. He retired aged 59. During retirement he took up long distance cycling and walking. He became the Audax UK Veteran Champion and with his wife completed many long distance walking paths in the UK and abroad.

Aged 70 he returned to academia to study Healthy Ageing at Guy's Campus, King's College London with his friend and colleague, Professor Stephen Harridge. Over the next 15 years their research, with other collaborators, produced many scientific papers. The research has been widely reported in media around the world and has been shown on BBC News, BBC Health and recorded by Channel 4 and Canadian TV.

In 2018, Norman was included on *The Sunday Times'* Alternate Rich List. At 84 he continues to carry out research on how to age well and wisely and is still cycling.

To June – for love, intellectual support
and advice; you never wavered.
Thanks to Nicole and Gavin, the next generation.
To Rita, at the forefront of our generation.

The Lazarus Strategy

How to Age Well and Wisely

Dr Norman Lazarus

First published in Great Britain in 2020 by Yellow Kite
An imprint of Hodder & Stoughton
An Hachette UK company

A CIP catalogue record for this title is available from the British Library

Trade Paperback ISBN 978 1 529 37669 2
eBook ISBN 978 1 529 37899 3

Typeset in Sabon MT by Hewer Text UK Ltd, Edinburgh
Printed and bound in Great Britain by Clays Ltd, Elcograf S.p.A.

Hodder & Stoughton policy is to use papers that are natural, renewable and recyclable products and made from wood grown in sustainable forests. The logging and manufacturing processes are expected to conform to the environmental regulations of the country of origin.

Yellow Kite
Hodder & Stoughton Ltd
Carmelite House
50 Victoria Embankment
London EC4Y 0DZ

www.yellowkitebooks.co.uk

Contents

CHAPTER ONE

Let Us Begin by Stating Our Purpose

'He not busy being born is busy dying.'

Bob Dylan

A moment ago, you decided to open this book. Life is full of these little decisions. Perhaps you are wondering what to do next.

Here's one good reason you should continue reading: this book is about you. Or rather, it is about you and me and everybody else, and the decisions we make and fail to make.

I am a healthy 84-year-old man looking both backwards and forwards, not only at the passage of my years but also at the lives of some of my past and present friends. I am saddened to observe the many opportunities they have squandered to live healthy, happy, productive and medication-free lives.

In this book I speak chiefly to those people who would, I believe, derive immediate benefit from a change in lifestyle. That means readers primarily aged between 50 and 70, give or take a few years, who are not being treated for any major illnesses. With certain reservations, I am also speaking to those who are in their 80s and beyond. Sadly, the vast majority of people my age are living under continual medical care. Why this is so will become clear.

The language of medical research tends to be written dispassionately, as though the findings do not relate in the slightest to the lives of the researchers. In contrast, this book makes no pretence of being a purely academic discussion. I will discuss the attitudes of many scientists and health professionals to healthy ageing based on my own research and personal perspective, which may reflect your own.

What is this book about?

It is a composite of my experiences as a research worker into healthy ageing, plus my experiences as a medical doctor and as a dedicated (ageing) exerciser. What emerges from this mixture is the finding that there seems to be an overwhelming consensus in the medical and scientific community that ageing is a 'disease'. An unusual disease, that is, in that one hundred per cent of the population will end up suffering from this 'disease'. Not a very sunny prospect.

I have a major problem with a view that inevitably links ageing to a disease. My contrary view is firmly founded on the research of my colleagues and on the large body of evidence that has existed in scientific literature for many years. This view states that the illnesses of ageing are strongly associated with behavioural and lifestyle choices. They are not the product of the inherent ageing process.

This book charts and follows the reasoning behind the conviction that ageing is not a disease. Come with me as I review the key research and let me unravel its meaning from the perspective of a healthy 84-year-old. I hope you will enjoy the journey, and at the same time uncover and understand your biological heritage.

What is this book not about?

This book is *not* about medical treatments. Nor is it an in-depth examination of what is happening to your various organs and systems as you age. I abhor that approach. We, the ageing, are not collections of medical specialties. We are the sum of all our parts. Read on and you will understand why I think the piecemeal approach is entirely counterproductive.

Making decisions based on facts

Everyone of my age, or thereabouts, wishes for self-reliance and independence. How to preserve these two attributes lies at the heart of this book. It is also a rallying call, a haka, a war cry; and a blast at those who rule the research purses and those who advise and treat us.

While I will offer a great deal of practical, sensible information, I do not aim to be prescriptive. People beyond a certain age have undoubtedly earned the right to make their own decisions. We do not enjoy being lectured to. However, there is an enormous difference in making decisions that are factually based and those founded on personal judgements.

The Lazarus Strategy

I aim to influence your decisions by placing before you all that I have learned, hoping that what you read will fit with your likes and dislikes, your habits and experiences, and so influence which road you decide to take. Remember that as age creeps along the decisions you take will not only affect you, but like ripples in a pond, the consequences will inevitably affect those near and dear to you.

Another reason for my opting to speak primarily to readers aged 50 and over is learned from my own experiences. The ageing process is quite sneaky. My apology to Bob Dylan, but to me his lyric suggests that dying begins as soon as physical development stops. This means that the process ending in death starts, for most of us, at around 18 years of age. The rate of change is gentle for the first four or five decades. This gradual decay is scarcely perceived, barely acknowledged and almost never acted upon. It is a slow process that does not appear to impinge on the lifestyle of the young. Perversely, the more inactive we are, the less the decline will be noticed. If you do not demand much from your body, the limited demands can be easily accommodated, even by a diminishing physiology. All will seem right with the world.

This perception is completely wrong. Mortality is a blind spot in us. It is only in the sixth or seventh decade of life that this view becomes severely challenged by reality.

So, if you are a young person perusing this book – and by that I mean anyone under 50 – it will make the perusal worthwhile if you have, despite your youth, reached a stage where you feel a change is necessary in order to reinvigorate the decades remaining to you. I have stood where you stand now and also thought that the future was far beyond the horizon.

It is not. Do everything you can so that you are able to enjoy the riches you hold for as long as possible.

To those readers in their sixties, seventies or eighties and beyond who are already leading active lives, then congratulations! This book is dedicated to you more than anyone else. Perhaps it will inspire you to maintain your vigour in the coming years.

If, on the other hand, you are among the majority whose lifestyle is not conducive to healthy living, then I hope you are reading this because you have decided enough is enough and you need to change before you become a source of income for the pharmaceutical industry. Keep on the path you have chosen and statistics show that most people after the age of 65 can, on average, expect to spend half of their later years with a life-limiting health condition (Age UK 2019). At current average life expectancy, that is about 10 years. This future holds a long period of morbidity sustained by non-curative drugs and the strong possibility of becoming an unnecessary burden on your children, family, friends and country.

The good news is that there is still time to change course for a better ending. Continue reading and I will show you how.

CHAPTER TWO

A Glimpse into Our Futures

The World Health Organization (WHO) predicts that by 2020 – that is now – the number of people aged 60 and over will outnumber children aged 5 and younger. By 2050 there will be two billion people over 60, with 434 million of those over 80 years of age (WHO 2018). These are remarkable figures with serious implications.

A few months ago, I sat on a committee comprised of the good and worthy that was specifically set up to discuss this approaching increase in octogenarians and the problems they were going to bring. The people sitting around the table were mostly in the 50 or so age bracket. Listening to their contributions to the discussion there was a schism between the thinking of the group and the upcoming problem. There appeared a complete lack of understanding that, in the time frame under discussion, the people sitting around the table would end up being part of the problem themselves. This group of middle-aged experts would seamlessly morph into the problem they were there to address. Their ageing futures would involve not only themselves but would also affect the lives of their spouses, children and grandchildren.

The sad part is that the problems would be avoided if they were to adopt the advice they were going to suggest for 'oldies'.

All too often these good-intentioned people interacting with the aged do not seem to grasp that they should first obey their own advice before handing it around. Doing so would provide examples of ageing to inspire confidence in those they are advising.

But of course, in all probability they would not.

This, you have guessed by now, is where we are today regarding both the people who give out treatment and the people they are treating.

What are the consequences of doing nothing?

Well, a collection of illnesses relating to bad lifestyle choices are waiting in the wings. There are at least twenty of them, most of which are preventable, including the following – cardiovascular disease (which can lead to heart attacks and strokes), pre-stroke hypertension or high blood pressure, type 2 diabetes, dementia, non-alcoholic liver disease, peripheral artery disease, certain cancers, not to mention overall frailty and infirmity. These diseases are serious conditions and can and do cause untold misery to those who have them. They may not, in every case, shorten our lives. However, this isn't necessarily good news. They will almost certainly turn a great many of us into incapacitated and unwelcome burdens on our families and the rest of society.

Let me be plain: we are going to get old. End of discussion. The decision of *how* we age however is in our hands. How we approach this inherent, ubiquitous, decremental process will determine whether we live an active, productive and independent old age.

Why so? Because research has shown that nearly all of these diseases are not due to the inherent ageing process but are largely the result of our behaviour and lifestyle choices. Now *that* is a wake-up call.

Some hurdles to overcome

Even when we recognise and accept this fact, we come up against two formidable obstacles.

Firstly, I must stress that we as a society have not yet begun to properly address those factors that are essential to age in a healthy way. Unfortunately, doctors and specialists have to spend much of their time, attention and effort studying and treating those of us bedevilled by the illnesses I have listed on p.8. Remember that these treatments are not cures, they are essentially holding operations – like sticking plasters covering the illness beneath. The more diseases you get, the more sticking plasters you will need.

The second obstacle is the different avenues and the welter of advice on how we can improve our odds of ageing well, making it difficult for many of us to separate the chaff from the wheat. How do I know what I should be doing? Every day I have to apply myself to the art of healthy ageing, continually trying to fathom what is essential and what is commercial propaganda. I make my decisions mostly on my years of experience and the results of my own research.

What have I concluded? 'Ageing' cannot be encompassed in

the misleading one-size-fits-all advice. Of course, we all age, but we can exert a great deal of influence over this inherent process. *How* we do that and regulate our ageing is the heartbeat of this book.

There are many erroneous beliefs about the ageing process. Some of these are propagated by our own lack of understanding of the link between ageing and behaviour, and others by many professionals not yet linking ageing disease with lifestyle and therefore concentrating their efforts in the provision of intensive and expensive medical interventions for diseases that are largely preventable. That is the tragedy.

These misunderstandings about the ageing process have also crept into the language that is commonly used to describe ageing. For example, contrary to popular belief, there is no way to 'slow down' or 'speed up' the ageing process itself. The inherent ageing process probably operates at the same rate in all of us. However, our lifestyles determine the state of our bodies. Present the ageing process with an active, fit body and that body will be able to better withstand the effects of the ageing process. Allow this same ageing process to interact with a sedentary, unfit and overweight body and that body will decline faster.

Adopting an active lifestyle appears to slow the ageing process, when in fact it is not the process itself that is slowed but its ability to *effect change*. That's why healthy, active seniors appear to have found a way to slow or stop the calendar.

How long have we known about the positive effects of physical activity?

As long ago as 1953, pioneering physicians Morris and Heady showed that coronary heart disease and type 2 diabetes were lower in people who did heavy work (Morris and Heady 1953). So the knowledge about the effect of exercise on the heart has been around for more than 60 years.

I would like to point out another important fact. Physical activity doesn't only benefit the heart. Just like ageing. Move and your heart responds, and then every other system responds to that movement. The earliest heart attack therapies were not only improving that organ but were simultaneously improving all the other problems that had been brought on by bad habits, such as high blood pressure and cholesterol.

Let us take a look at a simple analogy. Suppose that you are a robotic bread cutter who has been programmed to cut at a fixed speed and force – say five slices a minute – whatever the circumstances. You and your bread knife represent the ageing process, the bread is our body. You are given fresh bread to slice. The fresh soft bread represents a sedentary body and, in a minute, you cut your five slices. Next you are given a partially frozen loaf. This represents the exercised and honed body. Because you always cut at the same rate and force then the slices will take longer to cut through in the hardened bread. You will cut fewer slices in the same time. It is the bread – the body – which has changed but not the cutting which represents the inherent, regulated ageing process. It is much easier for the ageing process to attack the softened body than the fit and healthy body which provides more resistance.

Say hello to *'not ill but not healthy'*

Let me now introduce another new and very important concept into the ageing saga: the concept of *'not ill but not healthy'*. This term describes most of us, and particularly young sedentary people that you or I know. Despite the fact that their lifestyles are unhealthy they appear not to suffer any ill effects. Of course, this is just an impression gained from not knowing what is going on inside them. If these people were subject to a medical test, say to determine the amount of fat or perhaps cholesterol in the blood, many would show high levels. These high levels will not cause ill health now, because they are still at an early stage and to some extent protected by the resilience of the young bodies. But this high blood fat clearly shows that although they are not ill, they are certainly not healthy. Wait a few years and those fat levels will produce nasty effects, like heart disease or diseases of the arteries. I mention fat in the blood but I could also have chosen the immune system. As you age, analysis of your circulating immune cells would begin to show less ability to withstand infections (Duggal et al 2018). There are other measures which would also indicate that under the skin all was not right. Remember the phrase *'not ill but not healthy'*, we will be visiting it again and again.

For the reasons set out above, we can gather that it is imperative that all people, and particularly older people, should be active – but they can only achieve this by adopting the lifestyle that fosters that aim. Healthy living does not mean living forever. At school I was fond of the poet, William Wordsworth. Recently I re-read his 'Ode on Intimations of Immortality'.

Now, at my age, I find the poem too rosy-coloured. Over the past decade or so I have been having my own intimations, not of immortality but of mortality. I am not pessimistic about my inevitable demise; I am a doctor and I have long come to terms with a limited lifespan. I do not wish to live forever. I do not wish to halt time, with shades of the futility of King Canute ordering the tide to do his bidding, but I do intend to do all I can to mitigate the effects of the ageing process by whatever means available.

I seek a reasonable period of good health and a useful and productive engagement with the unexpected events life has to offer. I want to show that an active, productive old age is within the compass of most of us. The pathway to accomplish this has already been defined. I map it out in these pages so that you too can travel the optimum ageing route.

CHAPTER THREE

Who is Speaking to You?

I was raised on a small coal mine in the Orange Free State in South Africa. My father owned a general store at the mining company's behest. The one structure that remains fixed in my memory was the local school that was set up for both the miners' children and others who came from nearby farms. It consisted of one room, and in that one room children aged five to eleven or so were accommodated. Most came from poor families and many travelled to school via horse and cart, in some cases barefoot. There was a single teacher. The year was 1940, the Second World War was in full swing. Wartime, even when experienced on the fringes, was not a period of prosperity.

I contracted chicken pox at around that time and I remember during my recovery that the doctor held my wrist and looked at his pocket watch for what seemed a very long time. On recovery I designated my sister as a patient recovering from chicken pox. I needed a pocket watch. None was available, so using my initiative I cut the straps off my mother's expensive wrist watch given to her as a wedding anniversary present. I was now the complete doctor. On the discovery of the strapless watch, my nascent medical career followed the same spectacular short life of a Roman Candle firework.

The Lazarus Strategy

Retrospectively, it is difficult to know whether this first foray into medicine held the seeds of my future medical career but, like her watch, my mother's influence permeates through my early life like a meandering river on a flood plain. In our general store she presided over the area containing patent medicines. Their ingredients were of uncertain origin and most of the medical claims definitely bogus. In the absence of any medical presence my mother, over time, became the go-to person for medical advice. She had her own list of remedial specifics gleaned, no doubt, from taking care of three children. Her significant following suggested that she must have been reasonably successful, however, woe betide the unwary or the foolhardy who mentioned constipation. This word awoke the Sauron of the medicinal cupboard: Castor oil, and its Nazgûl-like helper – the dreaded tablespoon. There was no escape. The two wreaked vengeance on any blockage with the force of a fire hose on a matchbox.

Maybe it is this role of 'wise woman' that made the most lasting impression on my young mind. Oscar Wilde wrote that: 'All women become like their mothers. That is their tragedy. No man does. And that is his.' I appear to have escaped my tragedy.

Because of this association with remedies and their effects, I have always been fascinated by the mechanism of the action of drugs. While at medical school I volunteered, with my friend, to be a guinea pig in an experiment of which the aim was to determine the effect of whisky on kidney excretion. The free whisky might also have had an influence on the thinking of a young medical student. Every four hours we were required to drink a volume of whisky.

However, whisky first thing in the morning was awful and sitting around having to drink every four hours kept us in a state of semi-intoxication, with the result that I could not concentrate to read or do anything useful. We got out after a week. As far as I know, the results of this great research were lost. Another Roman Candle.

After my medical degree, I left my then benighted country and eventually arrived in the USA where I spent 10 years continuing both my medical and scientific studies. My wife and I and two children finally settled in the UK in 1970. My interest in experimental inquiry followed me, or perhaps pushed me, into joining the Wellcome Foundation, the pharmaceutical arm of the Wellcome Trust. Here I was involved in drug research on diabetes, among other diseases. Discovering a worthwhile drug that makes a significant difference in people's lives is on a par with winning a Nobel Prize. Sadly, I was not successful. I later left the Foundation and joined the Department of Health as Head of the Food Section. I then spent quite a lot of time following Health Ministers around, keeping them on track when they were asked awkward questions about the connection between food, chemicals and well-being.

At the age of 50 I was fat. My mother came from a family that tends to be overweight and it took a focused mind to escape from this family heritage of overeating to control my weight. Later, after I retired at 59, I began long-distance cycling in a big way. I found the lonesomeness of riding by myself suited my independent personality while the long distance fitted with my cycling capability. I was proficient enough to become the UK Veteran Champion in 2001, when I was 66 years old. Re-reading this paragraph, I am conscious that behind these bland statements lies a volcano of effort which I will detail later.

Based on what I had learned in medical school, I awaited with some trepidation the changes that take place in an ageing body. I expected there to be no more riding up hills, and to be taking pills to treat some illness or other that I would certainly develop. The list of possible disabilities is long. There are also the many other irritating problems that people of advanced age experience, such as not being able to get out of a chair, or not having the strength to open jars.

To my surprise, by the age of 65 none of these predictions had materialised in me, or in my cycling companions of a similar age.

Why were we different? Granted, not every cyclist was immune to the ills of ageing. And the differences in humans ensure that it is impossible to make predictions that are one hundred per cent correct. But most of my colleagues seemed to be getting along just fine.

I suspected that our lifestyles might be a major contributor to our sustained healthy physiologies so I visited at least four university departments asking for a chance to explore this idea. However, research into healthy ageing was not high on anyone's agenda. The subject is 'off' the medical curriculum. Also, the idea that exercise may be crucial to a healthy physiology had not yet penetrated many levels of medicine. 'Not ill but not healthy' was not yet a widely disseminated concept. It still isn't.

Finally, Professor Harridge at Kings College London

agreed to undertake a joint co-operative venture exploring ideas on the interaction between healthy ageing and exercise. The joint research that Professor Harridge and I conducted (and continue to pursue) has involved scientists at other UK universities and yielded valuable insights into the human ageing processes. The interpretation of these results as expressed in this book rests entirely on me, of which more later.

I and my long distance cycling friends became guinea pigs. I guess you need to be a certain type of person to want tubes stuck up your nose while you cycle to test your lungs' ability to expand while you breathe, or to have a biopsy taken from your muscle tissue. I never ask volunteers to suffer any procedure that I have not done to myself. There is nothing more satisfying than to combine one's experiences and learning at a late stage of life. I am able to relate my cycling ability to changes that are taking place in my physiology. I find this strangely fascinating.

A few years ago, *The Sunday Times* newspaper called me 'the octogenarian professor who holds the secret of eternal youth' (*The Sunday Times* 2018). Looking at myself in the mirror, this came as news to me.

When I was a young intern on my first day in the wards, I encountered two examples of mindless recalcitrance regarding health decisions that have never left me. The first was a young man, aged 35 or so, who sat on his bed smoking. This was in 1960 and no-smoking policies had not yet been introduced. (The Americans published the first reports on lung cancer associations with smoking in late 1950.) The patient was suffering from a medical condition called Buerger's disease which affects circulation in certain arteries, usually in the arms and legs, resulting in a lack of oxygen to the extremities. Lack of oxygen results in gangrene which is inevitably followed by surgery and amputation. I knew that nicotine constricts damaged blood vessels and the patient knew this also. In fact, by the time I met him he had already lost all of his fingers and was holding his cigarette between thumb and palm. Still he refused to give up smoking. I hope we agree that this was not a rational decision.

The second instance was, to my young mind, even more difficult to observe. A child born to a pair of religious fundamentalists required a blood transfusion in order to live. The parents refused to give permission, all the while loudly praying to God. They refused the treatment because the child's imminent death, they believed, was God's will. The child died. To me, neither of these decisions was explicable. But I still had a lot to learn. Problems that require difficult decisions are not solved by irrational decisions.

I am acutely aware that decisions relating to a change of lifestyle are difficult to make and to implement. I know there are no easy ways to lose weight, just as there are no easy ways to become active. After 30 years of weight loss I still have an internal dialogue that at times becomes a struggle. Logic instructs me not to eat too much, while at the same time the siren calls of pleasurable eating lure me towards danger. As we get older, these decisions become not easier but more difficult.

This book figuratively and literally opens another chapter in my life. It represents both a synthesis of the experimental results that I and my colleagues have found, as well as an analysis of what experience has taught me about the art of healthy living. I hope that this distillation will make it easier for you to adopt and then adhere to those principles that are the foundations of a medication-free life.

CHAPTER FOUR

Is Ageing a Disease?

I am incensed by the labelling of ageing as a disease.

The National Institute for Health Research recently published a report stating that roughly half of those over 65 suffer from two or more medical conditions (Kingston, Comas-Herrera et al 2018). It predicted that, by 2035, this will rise to two-thirds. In that year, the report also projected that three-quarters of people over 75 will have two or more chronic diseases – among them high blood pressure, lung disease, cancer, type 2 diabetes and arthritis. People may live longer than they have in the past but they will be imprisoned and tormented by illness. These tragedies wait to be played out. The diseases lurk inside us, like phantoms in the wings, until the first whiff of incipient dysfunction allows them to take centre stage.

Some predictions are even more dire. The journal *Age and Ageing* just reported that the number of people living with four or more diseases will nearly double between the years of 2015 and 2035, at which point they will constitute nearly a fifth of the population (Kingston, Robinson et al 2018).

Somewhere in the midst of those frightening forecasts and statistics are our futures.

An alternate view of the diseases of ageing

For many years, a small group of physiologists, based mainly in Scandinavia and the US, have been questioning the perceived wisdom about the relationship between ageing and illness. About 15 years ago Pedersen and Saltin, two Scandinavian physiologists, showed that considerable evidence had already been accrued in the medical literature detailing the effects of exercise on a number of chronic diseases (Pedersen and Saltin 2006). The role of exercise was further detailed by Booth and his colleagues in 2012, and published in the *Journal of Comparative Physiology* under the title 'Lack of Exercise is a Major Cause of Chronic Disease' (Booth et al 2012). They laid out the relationship between modern lifestyles and exercise, particular in relation to the onset of diseases.

A few years later, Pedersen and Saltin followed up their 2006 report with an expanded version entitled 'Exercise as medicine, evidence for prescribing exercise in 26 chronic diseases' (Pedersen and Saltin 2015). The authors carried out a comprehensive search of published medical and scientific literature. They particularly looked at those studies where exercise had been included in research protocols. They found 26 diseases which could be positively affected by exercise.

Many of these 26 chronic diseases are directly associated with lifestyle choices in ageing and the paper describes the effect of exercise in the treatment of these diseases: which include psychiatric diseases (such as depression, anxiety, stress); dementia; metabolic diseases (obesity, high fat levels in the blood, type 2 diabetes); cardiovascular diseases and their effects (high blood pressure, heart attacks, heart failure, strokes and pain in the calves brought on by exercise); pulmonary diseases (emphysema, asthma); and musculo-skeletal disorders (osteo-arthritis, osteoporosis, back pain); and cancers (two cancers, colon and breast, have been linked to lifestyle). This is quite a formidable list. All have the power to severely incapacitate and kill us. And nearly all are within our power to avoid by changing to a more active lifestyle.

More recently, it has been suggested – by others as well as by my colleagues and myself – that most of the diseases mentioned above, the so-called diseases of ageing, should be grouped and labelled as 'Exercise Deficiency Diseases'. The idea behind this move is to emphasise the similarity of these diseases by linking their root cause to a major deficiency of exercise (Lazarus et al 2018). This term is only newly published and we must wait and see how the scientific establishment reacts to this new overall classification of those 'so-called' age-related diseases.

Healthspan and lifespan

Healthspan? Never ever heard of it.

Look around you. Every biological system on the planet, not only human biology, has evolved by incorporating death into the package. No, this is not a Faustian pact. Faust received knowledge and worldly pleasures in exchange for his soul. Here, you keep possession of your life in its entirety. However, what you have is all you are going to get. Would it be possible, sometime in the future, to unhook life from death?

The Lazarus Strategy

Immortality has been a dream since Sumerian times. The legend of Gilgamesh, a king from Uruk, relates how he set out in 2500 BC or so to find a way to escape dying. Four thousand years later we are no closer to that goal and I'm happy with that.

We must now, though, differentiate between how long we live and the state of our health during those later years. Will our ability to lead happy, productive lives extend to cover those increased years? In scientific jargon, will our healthspan equal our lifespan? It was Bette Davis and Bo Derek who were reported to have said that ageing ain't for sissies. Right on!

As things now stand, there is very little chance of our health keeping pace with our life for most of us. Sure, all of us are predicted to live longer, but because of our lifestyles most of us will have to look to modern medicine to keep us alive. There is no doubt that modern medicine can do this but there is also no doubt that the medicine that we take is not curative and it comes at a cost. I must emphasise again that I am addressing the medical problems that are initiated and influenced by our habits, and not other illnesses that strike regardless of how we conduct our lives.

We need to ask ourselves a very serious question: do we accept that in order to enjoy our longer lives we must adopt a lifestyle that allows our healthspan to flourish over those extra years? Or are we so entrenched in our opinions that we would rather live our final decades suffering the failure of multiple physiological systems, propped up by non-curative polypharmacy? Then I suggest we think again. The choice is stark and uncompromising. A tragedy reminiscent of the great Greek dramas is being played out before all our eyes.

Castles in the air

Another disastrous effect of the attitude that medicine is the solution to our ageing problems is that this removes personal responsibility for our conditions. Why should we bother to better ourselves when the NHS is there to provide us with our predestined hand-out? We now have a bizarre situation where people have bartered their healthspan for a compromised lifespan, based on a totally unrealistic expectation of the healing powers of medicine. Are you willing to overeat and risk getting type 2 diabetes knowing that the drugs used for treating this disease are not curative?

How did most of us fall into this lazy way of planning for our ageing years? Let's take a step back for an insight into our attitudes and examine the data on which this dark view appears to be based.

We know that the majority of the human population does insufficient physical activity and tends to overeat. Yet they go about their business unimpaired for the most part and not yet under medical treatment. Once again: *'not ill but not healthy'*.

It is from this population that scientists almost always select the subjects for ageing research. While these elderly human subjects may not be ill at the moment, their physiology is on a fast-downward roll. The data assembled from these experimental subjects will not represent healthy ageing. They reflect instead the early stages of a disease process caused by inactivity and poor diet. Ideas about ageing based on these findings will screw up any perceptions we may gain about ageing. I would go so far as to say that all this data – libraries full of it – should be placed in one big pile and moulded into a statue representing folly.

Does the following make sense to you?

Imagine an individual who has spent their life eating too much, spending a lot of time sitting in front of the television and abstaining from unnecessary physical activity. Around 4 a.m. one morning a pain suddenly develops in their chest or jaw or back. This heralds the onset of a heart attack. The patient is rushed to hospital and excellent treatment results in survival and recovery. There is probably a trillion-dollar industry supporting this protocol.

But hold on.

All of the behavioural factors that precipitated the heart attack are still in place. What measures must be taken to ensure that the episode doesn't happen a second time, with perhaps fatal results? Most of us already know the answer. The person must achieve and maintain an appropriate weight, eat wisely and do enough physical activity. Either that or do nothing and seek out more pills and medical procedures and then hope for the best, even as they descend into an even more diseased state.

I think we all see the irony of this situation. Society strongly tempts us to adopt a lifestyle that will lead, almost certainly, to disease. The disease strikes. The patient is hospitalised. After this acute phase, the medical profession finally decides to recommend physical activity and diet so that the patient's heart can be given ample opportunity to regain its normal functions. Is there a horse around here or has it already bolted?

I recently read an article reporting that exercise is as good as pharmaceuticals in lowering blood pressure (Web MD 2018). I was not sure whether to laugh or cry. What else is new? Here is another question for you: if medical treatment saves thousands of lives, how many lives do you think prevention could save? I believe millions.

A look into pharmaceutical approaches still to come

Statins have been widely used for decades now to prevent a particularly widespread disease of ageing: high cholesterol. About 1.75 million people in the UK now take statins (BBC News 2019). Their use represents, to my way of thinking, a rather eccentric view of pharmaceutics. If we follow the thinking behind statins, we find that the idea is to identify people with early changes in blood-fat indices, and then target them for medication. In other words, treat people who are *not ill but not healthy*. Indeed, with that kind of thinking about 80 per cent of the population could be eligible for one drug or another.

I think it is worth stating again: ageing is a phenomenon involving all systems of the body. The only known effective therapeutic intervention is physical activity, which has the necessary global effects to strike at the root causes of 'diseases of ageing'. Chemicals definitely do not have this global action.

CHAPTER FIVE

Who is Max and Why Does He Matter?

When we think of our bodies and their capacity for physical activity, Olympic athletes and world champions might come to mind. Indeed, they are prime examples of what is possible. While the vast majority of us ordinary humans cannot hope to equal their levels of exploits, we undoubtedly have within ourselves the capability to reach heights that are appropriate to our own individual abilities and physiques. This applies even though we might have short, stumpy legs and scrawny bodies; or other deviations from those superb athletic specimens.

Championship performance is a relatively new human endeavour, probably started as recently as 700 BCE, in Greece. Athletes as sporting heroes are an even more recent phenomenon. The panoply of excessive worship owes much to Much Wenlock, a village in Shropshire, England, whose inhabitants were responsible in 1850 for reviving the ancient Greek games that eventually morphed into the modern Olympics.

By contrast, the human genome with its physical capabilities was laid down at least 40,000 years ago. It is important not to transpose the capability of breathtaking athletic performances of a few to our own everyday bodies. We ordinary folk engage in physical activity not for medals but for the personal glory of a life well-lived.

To illustrate this point, imagine a gentle incline outside your home. You decide to walk up it, despite not having done much exercise in the past few decades. The first time you try, you're out of breath. Yet you persist. Over time you become less breathless. Within a month, you are jogging up the slope. If you continue you could eventually run it. Your performance is nowhere near international competitive class but it may be the best that your particular physiology can manage.

Who knows? You may eventually end up running two marathons back to back or cycling a hundred miles. These feats do not require outstanding prowess. Only persistence. Forget about competition, with yourself or anyone else. Your wonderful, individual physiological endowment is a sufficient passport to health.

A question, then: having jogged repeatedly up and down that slope outside your home, are you now fit?

Before exploring this question, I wish to make clear that I make no distinction between movement, physical activity and exercise here. I use them interchangeably in this book. I am aware that academics take enjoyment from differentiating between these words but I can see no great illumination that will accrue for our purposes.

Perhaps we can sharpen the question posed by asking: fit for what?

Let's backtrack and think about the behaviour of our forebears, early Homo sapiens. Both men and women needed to be active in order to feed themselves, either by hunting or gathering. Being insufficiently fit to hunt was a death sentence. When chasing prey, they would have to make an instinctive judgement as to whether the prize was worth the chase, as lions and cheetahs do. It would make no sense to pursue prey for an extended period unless the creature could provide sufficient sustenance to replenish the energy spent in hunting it, and then a little more for daily living afterward.

Of course, there was little chance that these hunters were obese or out of shape. There probably never was overeating on a large scale, and an overweight hunter would not have inspired group confidence. Physical activity was driven by the need to hunt and to be sufficiently healthy enough to continue living.

Let us now look into the question of fitness. Perhaps the most instructive method is by conducting a thought experiment. Let us pretend that I have invited you to attend a lecture at my university. There is only one requirement: every member of the audience must not be receiving any medical treatment. In other words, none of the audience has been diagnosed as ill.

Surprisingly for me, about 300 people turn up. My talks must be improving. They are a mix of men and women and of young and old. I decide that this is an opportunity to enquire into their health.

I start with a simple question: 'How many of you have a body temperature of 37 degrees centigrade?' Everybody in the audience puts their hands up. OK, so that was not very helpful in trying to determine health differences. Another question: 'How many have a normal blood glucose level?' If they are a knowledgeable audience, and let us give them the benefit of the doubt, everyone will again put their hand up – and they do.

'Can you stroll around the block?' I ask. Once more, all hands rise. I am not having much success, am I? 'Can you do a full day's work?' All hands go up. I could carry on in this vein, eking out information from members of the audience but not learning anything that allows me to segregate them by health status. I need to change tack.

I now start from a different perspective. Ageing affects every system in our bodies and everybody in the audience is ageing. What questions can I pose that will provide the best answers about total physiological function? As we have stated, our evolutionary history tells us that healthy humans were hunters. Obviously, I cannot test the fitness of my audience by sending them all out to hunt in this university precinct. The only prey available would be unfit students and faculty.

What I can do, however, is to engage the audience in some activity that is analogous to hunting. I am going to ask them to run up a few flights of stairs. This exercise will certainly involve most of their important physiological systems. They will need to navigate, balance and use muscles, heart, lungs and sight. Running up a staircase is quite a complex activity.

Watching the audience tackle this task, we begin at last to observe differences. Some climb like gazelles; others struggle, huff and puff; while still others trudge slowly but resolutely. Let us immediately segregate the huffers and puffers. We discover that they are both young and old and all non-exercisers. This group has a good laugh at their inabilities and off they go for a snack. After all, what does it matter if they cannot climb a few stairs? Getting winded is not life-threatening. What does climbing stairs matter in a world of escalators and lifts?

Ah! But it matters greatly. Let's have a peek at the gazelles in the group.

Again, we find a mix of ages. All these are exercisers. They have adhered to their evolutionary heritage. But does this matter to their daily lives? How does their physiological fitness for running upstairs benefit them as they go about their business?

Before being able to answer these questions, we need to meet Max.

Introducing Max

During life, we will meet different versions of Max. There is Max(imus) the Roman general, Max Headroom, Max Speed and, finally, our Max. Our Max usually signs in as VO_2 Max. V is for volume, O_2 for oxygen, and Max is for maximum. Translated into English, this is shorthand for the amount of oxygen that a subject breathes when doing exercise at maximum effort. Generally, this is measured by having subjects run on a treadmill or ride an exercise bike to the point of exhaustion, or perhaps being whipped to climb stairs as quickly as possible. (Not really.)

We are ready to return to our audience members. We administer a VO_2 Max test to the huffers and puffers and to the gazelles. What do we find? The first predictable observation is that all the huffers and puffers have lower VO_2 Max levels than their exercising counterparts of similar age.

Remember, in this group we have people as young as 30 and as old as . . . well, old.

Now we'll test the gazelles of all ages, only to find to our surprise that the exercising oldies, those 60 or even beyond, can have a better VO_2 Max value than the 30-year-old huffers. There is a 30-year-age gap here. This finding is serious stuff.

We started out with everyone in the lecture being classified as 'not ill', as if everyone is of similar robust health, but now we discover that they occupy very different levels of physiological health. We see that as a result of levels of exercise, a 60-year-old can demonstrate better all-round physical ability than someone half their age. You might wonder why this is relevant. Both the 30-year-old and the 60-year-old manage to go to work, have families, have not been to the doctor lately, seem fully functional and enjoy life. What is the big deal?

Think about it again and you realise that to move oxygen efficiently to all parts of the body for effective exercise, you need good lungs and heart, a healthy circulatory system of arteries and veins doing their job, a well-maintained musculature, and a properly operating nervous system which allows your body to manage movement. In short, the ability to deliver oxygen during exertion is a fantastic measure of many body functions and of overall biological health.

Low VO_2 Max levels have been found to be indicators of general mortality rates for at least 40 years, with the first publications describing how it reliably predicts all-cause mortality (Blair et al 1989). Those 30-year-olds with a low value may not presently be ill but they will be rapidly approaching a threshold level that, once passed, will herald illness.

So, the ability to supply the body with oxygen when exercising is probably the single best measure of physiological fitness, meaning health. *The single best measure.*

Despite all this – despite your knowledge of our evolutionary heritage and its link with physical activity, despite now knowing that 'not ill' does not imply excellent health – you may still be unconvinced that exercise is a necessity at this stage of your life. Perhaps you need further proof.

As I've said before, when many scientific researchers conduct studies, they tend to ignore older people who are active and healthy, mainly because there are so few of us (them). In 2018, however, the *Journal of Applied Physiology* published a paper titled 'Cardiovascular and Skeletal Muscle Health with Lifelong Exercise' (Gries et al 2018). Researchers at Ball State University in Muncie, Indiana in the US decided to study people aged 70 and over who had begun working out during the fitness fad of the 1970s but – unlike most of their peers – had kept at it, not for weeks or months, but for five decades.

Using newspaper advertisements, twenty-eight older males and females were found. Another group of subjects of the same approximate age but completely sedentary was also recruited (it was not terribly difficult finding those). And, finally, a third group of active people in their twenties was assembled.

They brought all three groups into the lab, had them exercise and tested their aerobic capacities – their VO_2 Max. Using tissue samples, their blood vessels and the enzymes contained in their muscles were also evaluated. The researchers focused on the cardiovascular system and muscle tissue because both, it was believed, inevitably decline with age. The researchers assumed that the young group would possess the greatest aerobic capacities and most robust muscles, with the lifelong exercisers being slightly weaker on both counts, and the older non-exercisers even worse.

That didn't exactly happen.

The active older subjects *did* have lower aerobic capacities than participants who were three decades younger. But the old exercisers were about *40 per cent better off* in this department than their inactive peers. In fact, those old exercisers had the same overall cardiovascular health as inactive people 30 years younger. Those non-exercisers will someday need to find robust young donor hearts to achieve that same feat.

And as if all these benefits were not enough, researchers also discovered that the muscle vessels and muscle chemicals of the older exercisers were all found to be well preserved and showed the same activity as younger people. Meaning, of course, that their muscles were dramatically healthier than those of their peers who had sat around doing little for the previous 30 years.

'In summary,' the study's authors wrote, 'lifelong aerobic exercise provided substantial benefits for VO_2 Max and skeletal muscle metabolic fitness among women and men in their eighth decade of life. The greater VO_2 Max in the [lifelong exerciser] cohort compared with [older, healthy non-exercisers] remarkably decreases the relative risk of mortality and provides for a large physiological reserve'.

Research conducted by our team at Kings College London examined the effects of exercise on a group of highly active older cyclists – males and females between the ages of 55 and

79 (Pollock et al 2015). These were not competitive athletes. They nearly all belonged to Audax, which is a non-competitive long-distance cycling club.

What was learned? Over a large range of indicators of physiological health, these cyclists, both men and women and age for age, had higher VO_2 Max levels, better muscle structure and physiology, and higher mental agility compared to people who were not active. On virtually every attribute for which they were measured, they scored significantly higher than non-exercisers of their same age. I am pleased to report that I was one of the volunteers studied and it is satisfying to record that my VO_2 Max was just fine. (If you want to know the value, it was 53 ml/min/kg. So there!)

An extra surprise finding was that these elderly cyclists' immune systems had been protected from ageing effects. Since the immune system protects us against a wide range of nasties, everything from colds to cancer, it was clear that the active elders were better prepared for life's vicissitudes (Duggal et al 2018).

Since the publication of these papers, more and more research has begun to tease out the differences between exercisers and non-exercisers. So far, almost all measures are superior in active people compared to their sedentary counterparts. It is becoming clear that our behaviour, our physical activity, affects not only our whole body but also all systems, including the cardiovascular, immune and nervous systems. The effects of being active trickle down to cells and even to the hormones circulating in the blood and tissues. A bonanza of bonuses.

A quick recap

What should we conclude from all of the above? As we've seen, the single-most important measure of overall health is reflected in VO_2 Max and its requirement of pulmonary ventilation and the ability to breathe effectively, the capacity of your cells to take up and utilise oxygen, blood flow into the muscles, cardiac output and blood volume, and sufficient haemoglobin to carry oxygen. Measuring VO_2 Max is like obtaining a status report on the state of your entire body. That is the reason VO_2 Max is recognised the world over as being the gold standard for predicting your likelihood of dying in any particular year (or not).

What if you don't feel able to identify with these aged lifelong exercisers? Would you agree that ordinary people can do extraordinary things? Consider how many marathons there are around the world and how many people take part. The vast majority are ordinary people, very similar to you and me except they wake up one morning and decide they need to test whether they can run a marathon. Why? Well, I suppose because, like the mountain, it is there. With proper training they can and do.

Every year there is a cycle ride from London to Brighton, a distance that is close to 100 kilometres. Guess what? About 20,000 people, no different from you or me, take part and complete the ride. Think of the hundreds of thousands of people who take part in a 5 kilometre park run each weekend. I could go on with other examples but that would be smashing a nut with a steamroller.

What is important is that nobody cares how fast you can run, swim or cycle. Nobody cares what event you join or what position you attain. It is taking part and finishing that is the reward in these events. Also, remember that all these people taking part are completely self-motivated. There are no prizes or laurel garlands awarded. Just self-satisfaction. We all have the necessary genes, the necessary muscles, the necessary brain power to take our bodies to whatever limits we can conjure up for ourselves. Elsewhere in this book, I enlarge upon our attitudes and emotional interactions with physical activity.

The message is clear. Just go out and move. If you have not previously been using your physiological gifts, you will be surprised by how quickly your body will respond. Muscles, lungs, heart and mind will figuratively love you.

In other words, you will begin to love yourself.

CHAPTER SIX

The Trinity That Influences How We Age

In 2015, the WHO defined healthy ageing as 'the process of developing and maintaining the functional ability that enables well-being in older age. This includes the functional capacity to use all mental and physical capacities and includes their ability to walk, think, see, hear and remember'.

That sounds simple, except that a very important aspect of ageing is conspicuously missing. The ageing process is not static. It is continually and continuously changing all our physiological *and* mental processes. The inevitable result of this process is a steady diminishment of nearly everything. Sorry to burst your bubble but it's true.

How then do we formulate a way of addressing this ever-changing landscape? How do we keep ourselves healthy? We do this by working on the three main factors that are crucial to our health.

It is tempting to use familiar but inappropriate analogies to describe these factors. It would be easy, for example, to talk about the 'three pillars' that sustain health. The problem is that pillars are a great metaphor to describe static situations but in no way can they be used as descriptors of the continually changing environment that epitomises the human ageing process.

A different, more fluid and amorphous analogy is called for and I am struck by the nomenclature of quantum electro dynamics, where both colour and flavour are used as descriptors. This use of the abstract qualities of familiar nouns to describe properties of a definite but also nebulous particle appeals to me. I will use colour to describe the three fundamentals of behaviour that underpin healthy human ageing in order to highlight their shifting and changing relationships to each other.

The Healthy Ageing Trinity

Colour Red: Exercise

Physical activity – affects all our physiological systems. Physical exercise is riveted to our DNA. To choose not to perform it is one of the worst health decisions any person can make.

Colour Blue: Food

Food – provides the energy for physical activity but also supplies the nutrition needed to replenish our bodies and allow us to expand and flourish. Too much eating, like too little physical activity, is a disaster.

Colour White: The Mind

The mind – a healthy, active mind oversees and monitors all aspects of living.

I wish to emphasise two very important points: firstly, each member of the trinity is inextricably bound to the other. Like the Three Musketeers they are 'all for one and one for all'. For example, enthusiasm can energise you but if you become downhearted or depressed, your state of mind affects your desire for physical activity and possibly your eating habits, even though nothing has changed in both systems. Similarly, if you exercise, you increase the blood flow to your brain and initiate thoughts that influence your willingness to exercise. Food supplies building blocks to all parts of the mind and body, and without this fuel for the regeneration of muscles you wouldn't have the energy to move.

Secondly, all three have body-wide effects. The investigation of all their integrating abilities cannot be studied by dividing the body into systems. That approach is based on a model designed for the investigation of disease and is wrongly used by the great majority of scientists to address ageing. Using this inappropriate, disease-centric approach, we lose sight of the global nature of the ageing process.

The trinity in hominoids

Let us revisit our hominoid ancestors. We agree that they must have been physically active people who gained sufficient food that they acquired through their ability to plan and orchestrate hunting and foraging. Those two critical issues – hunting and eating – were completely intertwined. It's easy to see why these two, now defined as physical activity and healthy nutrition, are still of the utmost importance to our existence.

The hominoids also needed active minds to hone and integrate their hunting processes and skills. The state of their minds was integral to their successful existence. The trinity of mind, body and nutrition needed to be in place in order for our distant ancestors to remain alive.

Today, the overwhelming majority of humans partake in little or no physical activity, and in addition overeat sufficiently to become either overweight or obese. Our trinity is not in cohesion. The outcome of this lack of cohesion will be one or more of the so-called diseases of ageing.

Painting a picture

These shifting relationships among the trinity also bring home the idea that throughout our ageing decades a continual adjustment must be made to the way we think, move and eat. Keep a flag of red, white and blue fluttering in the breeze in your mind as a constant reminder of our commitment to a healthy lifestyle.

Let me try to paint a picture of ageing using the three colours. Give your flag the dimensions of a metre square. That is the figurative size of your physiological heritage. Now, we need to add the three colours to the flag. We add red, blue and white and, for the convenience of this discussion, let us postulate that each occupies one third of the flag. The exact area each colour covers will depend on your particular physiology (but does not affect the analogy).

Now, as the ageing process exerts its effect the size of the flag begins to shrink. If you have been wise and living the 'correct' lifestyle, the shrinkage of the flag is coherent and synchronous. Although the areas of the colours shrink, they will keep the same ratio to each other. Retaining the same ratio is the crucial physiological regulation. The result is that all physiological processes are kept in concert. There is no disease process operating. The integrated trinity operating via their 'one for all and all for one' will ensure optimal ageing.

However, if you are sedentary, often overeat and are perhaps slightly downhearted then the three colours are certainly out of kilter. The area each colour covers and their ratios to each other will be different from that produced if you lived your best lifestyle. As the ageing process begins to shrink the flag, the coloured areas will also shrink, but now their ratios will not conform to that required for healthy ageing. Out of this disruption disaster awaits. Physiological process will fall out of synchronicity and the integrity of the shape of the flag is compromised. The disjointed trinity, adhering to their 'all for one and one for all', now hurry your body to misfortune.

Trinity #1: Exercise

Back to our ancestors

I have referred earlier to our human evolutionary history and how activity has been integrated into human physiology. Our knowledge about human behaviour in prehistoric times is a mixture of speculation and some fact. Genetically, we are a mixture of Neanderthal and early Homo sapiens DNA. About 1.5 per cent of us is Neanderthal. Reading some accounts of those distant times, it's easy to get the impression that there were only men around. It is safe to say that both men and women were needed to provide sufficient food.

There was clearly a close connection between being physically fit and surviving. Notice there is nothing here about athletic competition or challenging your cave neighbour to race to the nearest tree. No Neanderthal or early Homo would even contemplate going for the burn in this way.

Activity at the extreme is not essential for healthy ageing

Remember that performance in pursuit of a medal is a relatively modern human endeavour. It is important not to identify sporting heroes as the sole exemplars of our physical abilities. Athletes are obsessive about their sport and themselves and will happily devote their young lives to this obsession. Generally, they burn out within four decades. All this is admirable if you want or need a gold medal or indeed any medal but not so useful to those of us who do not conform to the required physical and mental attributes and are already well past our sell-by date.

Without doubt, these athletic champions are exceptional, but their achievements are available only to a very small percentage of the global population. We, the readers of this book, are of an age where we need to engage in an endurance activity that will carry us through all our remaining decades – that will mean more to us and our loved ones than a bucketful of medals.

Back to the here and now

So, maybe you are reading this book while enjoying your mid-afternoon snack. 'What has all this got to do with me?' you might wonder. Perhaps at school you were useless at sport and the last person picked for teams – all of which was embarrassing and degrading to one's ego. In addition, the last thing you wanted was to get sweaty all over. Since those far-off days you might not have bothered to do much physical activity apart from running for the bus/train occasionally.

It is this loss of movement or physical activity or 'exercise' that I am trying to change. Human evolution has endowed us with all the necessary ingredients to move. Movement is still integral to our physiological as well as our mental well-being. You may be loath to accept this truth but there it stands. We ignore it at our peril.

Getting on first name terms with the power of physical activity

Think back to that little incline that we mentioned previously and which you probably avoided because it made you huff and puff.

Today you decide you are going to stretch yourself. A new dawning. You will try to reach the top of the incline. Notice there is nothing here about competition, body size or shape. Those appendages at the ends of your legs are adapted for upright walking and you have set yourself a challenge. To quote Mao, 'A journey of a thousand miles starts with a single step.' Today is a beginning. Nobody apart from you cares what is happening.

You persist. Over the days you get less breathless. If you do not feel better after trying this activity for a week, I will stop eating porridge for a week. Never knock the effects of exercise. Even a slight improvement in your physiological well-being can make a dramatic improvement to your life. The results your efforts have produced are probably the best that your particular physiology can perform now. That is absolutely fine. You are allowing your body to use its evolutionary heritage. Who knows where this will take you. These feats do not require outstanding physiological athletic prowess. They only require the will. If you can walk, you are a prime candidate to begin.

Dog days

Still not convinced? Dog owners need to take their dogs out every day. Some dog owners do this purely to ensure that their dogs do not relieve themselves indoors. However, dogs need physical activity too. The creatures can be high, low, short-legged, long-legged, big-headed or small-headed but they love using what abilities they have. Well-exercised dogs have an aura about them. They are lean, happy and alert. They are ready for anything. Responsible pet owners accept these truths.

When it comes to basic physiological needs, what makes you think we are any different from our dogs? A study examining the functioning of dogs' hearts found that when dogs were exercised on a treadmill at 9 kilometres per hour for an hour twice a day for ten days, a chemical – nitrous oxide – that dilated their coronary arteries and allowed their hearts to get more oxygen was markedly increased (Sessa et al 1994). Ask your dog to take you out for walkies and more.

Trinity #2: Food

When I was a whippersnapper, arterial diseases were coming to the fore. It was post-World War Two, food rationing was over and diets were becoming more varied. With this diversity came excess saturated fat, and with this came diseases of the arteries such as atheroma. Our intake of food – both in terms of the types and the amounts – has changed dramatically over the years since then. An NHS report estimates that the average adult takes in 200 to 300 more calories than they need every day (NHS Eat Well). This might not sound much but over time it will have a massive impact.

Some easy arithmetic on gaining weight

Our body saves any energy that we don't burn. Nearly all that increase in calories will be stored in the body as fat. When we still endured famine or food scarcity, storing a little extra energy was a good idea. Today, food is all around us so that excessive energy storage is not necessary.

Let us do some simple maths. A kilogram of fat in a human contains about 7,700 calories. That is a huge amount of energy. Just how huge? If you are a 70-year-old woman, you should be eating about 1,600 calories a day. You will immediately realise that you could live for four to five days on that amount of calories without eating. We now go back to that increase of 300 calories daily that we don't need and multiply it by 365. This works out to about 110,000 calories in one year, more or less.

Galloping goblins! If we could live on one kilogram of fat for four days, how long could we survive on that spare 110,000 calories? I do not want to know the answer. If you are braver than I, divide 110,000 by 7,700 and you will discover how many added kilograms of fat us overeating Britons would pack in on a year of overeating.

I couldn't resist . . . The figure works out to 14 kilograms of pure body fat. If you continue to overeat to this extent you would eventually burst. OK, 300 calories is overeating on a major scale. Let's bring that figure down to a realistic figure – say, 125 calories added per day. The fat gain works out to 6 kilograms per year, which sounds bad enough.

Now, multiply that by 20 years. You can see the danger.

How much would you have to eat to ingest 125 calories?

What does 125 extra calories a day represent in food terms? That could be a coffee with cream, some added flavour like a caramel or hazelnut syrup, and a marshmallow thrown in. Half a sandwich would also do it. You can see how easy it is to add 125 calories to our daily intake. If you include it in drink form instead of food, it's almost imperceptible. I'm trying to prepare you for the tightrope you must walk to keep yourself at a desirable weight. An extra half a sandwich a day over time will equal obesity! Falling off the tightrope is as easy as eating a slice of your mother's apple pie.

There is an upside, however. If you are at your correct weight and then put on a half a kilogram or so, relatively minor adjustments to your intake will bring your weight back down. If you cut out a half-sandwich a day, and provided you do not backslide, then over time your weight will drop back down.

A stroll around a supermarket

Let me invite you to take a stroll through a supermarket with me. As we pass the biscuit shelves, you select a packet and ask yourself, 'Really, is one measly biscuit going to affect my daily calorie intake?' Luckily, there is an interesting bit of information on the label. It reads:

> *The energy in each biscuit in this packet equals 71 calories. This equates to 4% of an adult's reference intake.*

Quick, tell me, 'What is your total reference calories intake? If fat makes up 4 per cent of each biscuit how much will the biscuit contribute to your overall fat intake for the day?'

If you have had a biological education and are paying close attention to your food intake, you may score full marks. What if you have very little dietary or biological knowledge? How do you decide what to eat and how much? What about if you wish to lose weight? You may spend a lot of time trying to answer these questions. For me, life is too short.

The present labelling of food is a sop to the chattering classes. What you actually need is far removed from the information presented by a barely readable label. You need to teach yourself the basics about food and calories. I suggest consulting *The Composition of Foods* by McCance and Widdowson as a starter (McCance and Widdowsor 2014). That knowledge is a form of preventive medicine in itself. It is planning for the future.

More mixed messages about food

Let's keep on the subject of information on food for a moment or two longer. Are you one of those shoppers who read the literature that some supermarkets issue on healthy living?

In each supermarket magazine there will be a mixed message. One half will inform you of the great advantages of climbing aboard the various health-oriented food regimes (vegetarian, vegan, whole foods, etc.). The other half will be advertising a whole range of foods containing far too much salt, or too much added (and unnecessary) sugar, or too much fat, and so on. Are we being advised to buy these foods too? Should we stop by the confectionery counter also?

How about that lovely chocolate muffin at the bakery section? It contains about 500 calories and 20 grams of fat. That one bit of fluff could be about one-third of your recommended daily intake. Is that a bad thing?

Do you see what I mean about mixed messages? Most store policies seem to boil down to this: we, your friendly neighbourhood supermarket, will show off our healthy eating credentials in the magazine but also stock aisles of foods that are anything but healthy. Things we know should not be on our shelves or in your kitchens. Buyers beware.

The hospital café

I know what we can do – let's stroll into a hospital café set up for visitors and some patients. Surely the food here is carefully monitored. You must not have been in a hospital lately, or ever, as like the supermarket, this institution will feature those foods that generate maximum profits.

Also, if the hospital staff represents a true cross-section of the population, then many of the medical and ancillary staff – the ones giving out advice on healthy ageing – will be non-exercisers and probably overweight.

The private doctor's waiting room

I accompanied a friend recently to a doctor's appointment. There was, as usual, a television blaring at a volume just sufficiently annoying to raise my blood pressure in the waiting room. If the television had been off, the music would have been on. I wanted to scream. So much for me being able to relax in silence and concentrate on my friend's coming encounter.

There was also a coffee urn, and with the free coffee came as many biscuits as you wished. Each contained 100 calories and 40% fat. How could I possibly be expected to spend a half hour waiting without eating?

Consulting those books on diets

Our food environment is doing us no favours, so why don't we consult the experts on diet. What better place to start than in a bookshop? But which gurus do we follow? There are hundreds to choose from. Do we want the high protein diet, or an intermittent eating diet, or the diet for vegans or carnivores or carb-lovers or carb-haters? I assure you we are nowhere near the end of the dizzying selection available. It is depressing to find that people who should know better still do not appear to have figured out that the world is populated by diverse cultures and that dietary habits are not necessarily rooted in their own dietary history. Which book will cure you of your eating habits? Do not look to a book. Look to yourself. By now you will have begun to realise that by adopting the trinity, you can have a crucial say in the quality of your ageing years.

Some thoughts about food and society

Food has connotations way beyond its nutritional value. Religions use it to foster group identity in their adherents. Perhaps the avoidance of some animals for food is the first indication of a nudge towards vegetarianism. However, at times I am struck with the notion that the eating of vegetables may be being raised to similar religious heights.

Food is at the centre of our social gatherings. You try visiting someone and getting away without eating. The ubiquity of food – the fact that it is virtually everywhere that we are, more so than ever before – is a big problem for those of us trying to control our weight.

Pavlov's dog

Ivan Petrovich Pavlov, of course, was the Russian scientist working at the beginning of the 20th century. He showed that dogs could be conditioned by teaching them to associate a ringing bell with food. The sound alone then caused the dogs to salivate in anticipation of a meal.

When I was in my early 50s I was about 12 kilograms overweight. There were many reasons for eating too much but a big one was because I was bored. I ate at work, at home, on aeroplanes, train journeys, you name it. I was the human equivalent of Pavlov's dog and I had done it without help: get bored, relieve the boredom by eating. Soon, every time I was bored or *thought* I was bored, I ate. If you are overweight it is almost certain you have your own self-taught cues to eat. What are your cues to eat?

Finding the drive to eat less

Put all the above together and you can see that losing weight requires more than just eating less. Sure, that is the basic mechanism, but the drive and the method must come from within. You can try to fake it but the rigour needed to persevere will find you out. Your mind will tell you that you need food even though you can see and feel your energy storage unit – your body fat. You need to decide which part of your brain is going to be in control. That part urging you to eat or the part where logic resides. This requires effort. I know because, after 40 years, I still need to use judgement and control. I cope in one of two ways, depending on my mood. When hungry I might go for a short stroll, or concentrate on the book I happen to be reading to distract myself. Or perhaps I will test myself to see if I can meet the hunger feelings straight on and resist them. There is no easy way to lose weight. Keeping it off is even harder. You must develop the courage, fortitude and will to sustain your new lifestyle and not revert to your previous eating habits. Your mind must stand, like the fortress of Malta, a bastion against the hordes of invading food.

If a sure-fire curative treatment exists then it is obvious to everybody

When I was a young medical doctor in South Africa, tuberculosis and its ensuing conditions were still rife. There were sanatoria set up in cities and in pleasant country environments to accommodate the patients. Fresh air was supposed to be curative. The patients were subjected to all kinds of bizarre treatments (like collapsing an affected lung because medical theory held that a collapsed lung was a resting lung).

The first antibiotic to be effective against tuberculosis, streptomycin, was discovered in 1943 by Selman Waksman. Streptomycin changed the landscape of tuberculosis. The point I'm trying to make is that if there is an effective treatment, then everybody takes that cure. When you find so many weight loss gurus in the world, you must assume that no one has a proper handle on how to lose weight *and* keep it off. The popular statistic is that about 95 per cent of people who lose weight, *whatever the dietary regime*, will regain it – and then some. There is a large, vulnerable population out there wishing for a body other than the one they have.

Trinity #3: The Mind

Does the mind play a part in human ageing?

This question is, to a large extent, rhetorical. After a certain age, we can all describe some of the tricks that the mind starts playing on us. You see someone you have met before but you cannot recall their name. You put your car keys in a safe place and then forget the location. You walk into a room and cannot quite remember why you are there. These tricks have profound effects on an old person's view of their life and of the world.

If I had to construct a hierarchy of the trinity, I would tend to place the mind slightly above the other two members of the trinity. However, I do not feel that this hierarchy in any way really changes the way that we need to approach and understand the inter-relationship of the trinity and healthy ageing.

Some effects of exercise on mind functions

The benefit of exercise on brain function is well recognised, although not generally known by the public. I quote from an article in the *Journal of the American Geriatrics Society* (de Andrade et al 2013): 'These exercise effects are extended to individuals with, or at risk of, dementia and other age-related neurodegenerative disorders. Accordingly, although extensive studies are required to fully understand the mechanisms by which physical exercise procures beneficial effects, data suggests the relevance of integrating physical exercise in the prevention and/or cure of Alzheimer's disease, disease whose incidence is predicted to increase in the future'.

In our study of mature cyclists at King's College London (Pollock et al 2015), my colleagues and I found that the cyclists' efficiency at information processing declined with age at a similar rate (5–10 per cent per decade) to that seen in the general population. The test involved the subject being given an A4 sheet of paper with 65 letters on it and having one minute to cross out as many 'P' and 'W' letters as possible. Processing speed was determined by the total number of letters scanned and efficiency by the percentage of target letters identified. OK, so exercise may not be a cure-all, but it is still better than any drug you can name or swallow.

Let us agree that the inherent ageing process is controlling all mind and body functions. We must stop thinking in terms of separate systems when we consider ageing. There is no division, they are irrevocably linked. We do not live out our lives as separate systems.

Mind games?

You will now have twigged that asking what and how much exercise one needs to do in order to keep the brain up to scratch is falling back into thinking that the brain is somehow separate from the body. It is not possible to do brain-specific activity any more than it is possible to do activity that is specific to only the heart, or just the musculature. The effects of most exercise spill over into many other systems.

Of course, you can sit in a chair or lie in bed and do mental activity such as puzzles and crosswords but all this will do is increase your proficiency at the small task at hand. This mental exercise has as much beneficial effect as a leisurely stroll around the block – meaning not much.

While engaging in physical activity, your mind is directing all operations. That is its job. As you improve your physical abilities, your mind can begin to think of attacking different and perhaps more difficult challenges in life that you thought were beyond you. These major executive decisions are the decisions on which the brain thrives.

Attitudes

One of the frustrations visited on the elderly is being advised by people twenty or thirty years younger who do not them-selves adhere to the advice being given or to the trinity. Are us oldies supposed to be impressed by the old chestnut of 'do as I say, not as I do'?

The other assumption here is that advisers, about three to four decades younger than the people they are addressing, have no trouble understanding what is going on in the mind and body of a 70-year-old. If we truly followed this thinking, we might as well also hire 16-year-olds to advise 30- or 40-year-old patients. Indeed, they appear to understand so well that they are of the opinion that oldies, even healthy oldies, need not be consulted whenever they launch the next hare-brained scheme to control the lives of the elder generation. What could intelligent, educated people who have already faced these challenges contribute to the discussion? Apparently very little! We suggest the initiators of programmes look to their own behaviour before they look towards ours.

CHAPTER SEVEN

Putting Activity into Practice

Apparently, 17 per cent of the British population belong to a gym, and 13 per cent of us exercise regularly (Statista 2019). (Please keep in mind that this latter figure is self-reported and should be viewed with scepticism.) My own guess is that only between 20 and 30 per cent of us get any significant exercise whatsoever. Around 23 per cent of all Americans currently meet their national physical activity guidelines, according to a new report (Blackwell and Clarke 2018).

There is all that knowledge out there about activity but we do not seem to get the message. One thing is clear: just telling people what to do, even when it's for their own good, is close to useless. Nobody knows better than I what a dismal failure all that well-intentioned scientific research has been. Most of what I study is reflected in these dismal figures.

Take a look at me

I was about 54 years old and taking a walking holiday in Switzerland. Despite my knowledge about health, I was eating much too much and about 12–15 kilograms overweight. Sitting down to dinner with my wife in a hotel in Saanen, I happened to glance down and saw my stomach bulging over my belt. I had seen that before but somehow the significance had passed me by. There and then, and coincident with the arrival of the soup course, I decided enough was enough. I was going to get into shape. I laid down the spoon and refolded the napkin. These simple gestures were symbolic of my change of heart and my determination. When we arrived home, I began to explore strategies for how to lose weight. That was how I made my decision. No long period of stuttering and muttering, no soul searching. Just go and do it.

The mind over matter

To change your life, you must start with the mind. You must know deep within yourself that you wish to embark on this odyssey. It is no good kidding yourself. Imagine a sporting team huddled on a playing field, its members trying to psyche themselves into believing that they could defeat an opponent they knew in their hearts was superior. Doubt is a very peculiar emotion and it tends to get more insistent as we age.

I remember when it first hit me. I had been cycling for 20 years without a worry about terrain or distance. At about 75 years of age a little voice asked me whether the route was too difficult for me. Never had I heard that voice before. I suspect that it was linked to the changes taking place as I aged. Perhaps it was a warning signal from my mind to my body.

You are embarking on a course here which will extend to your last years. You may hear that voice some day too but don't permit it to control you.

Physical activity

If you are now, as I once was, somewhere between 50 and 70 years of age, maybe older, yet hopping up and down and ready to go, then you may be expecting me to prescribe a physical activity programme that will transport you to superior health.

No, I reply. I will not. I cannot.

In my opinion, that is the wrong way to approach your needs. Let us start somewhere else. What are your aims? What mental picture have you formed about the goals you wish to attain? What are you going to do with your new-found physical and mental abilities? Mindless physical activity will simply wear you down or you will lose the will to continue.

When I started all those many years ago, I sat myself down for a good question and answer session. Did I *enjoy* – notice the word – running? No. Scrap running. Swimming? Yes, swimming was possible but coming to England from a sunny country, South Africa, I did not like the atmosphere of an indoor swimming pool. Plus, I did not wish to do an activity that confined me indoors for long periods. How about walking? I love walking with my wife but I was looking for a new challenge.

Cycling? Let me think about that. Cycling is generally a summer sport in the UK so I'd have to be prepared to go out in all conditions to keep myself active all year-round. If you

cycle, you get wet. So what? I knew that my skin was imper-
meable and that clothes can be dried. The advantages of
cycling were that it fitted in with my personality. I could do
it by myself, I could go at a pace that suited me, and I liked
the idea of covering long distances in the open air at a speed
I regard as human.

June, my wife, on the other hand, loves the outdoors and
walking. That was, and still is, her central physical activity.
Together we expanded our walking horizons, and by including bird watching, plant hunting and butterflies, made every
walk into not only a physical but also a mental adventure.

How about yourself?

Start by choosing a physical activity that you will enjoy.
Maybe it's a group sport like aerobics, or something you can
do with a partner, like tennis, badminton or dancing. An
acquaintance moved a cross-trainer apparatus into the garage
and used that – a good solution for someone who wishes to go
solo. The list is endless. Never lose sight of the enjoyment
aspect. If you do not get your mind into the correct mood,
your physical activity will just become a burden.

Your horizons

I am going to assume that most of you reading this book will
not have taken much physical activity lately. So, what horizons are open to you?

Putting Activity into Practice

I think that at 50, you are in your prime. By now you should know yourself and your limitations. You will have sufficient experience of life to have faced problems and solved them. You are probably as near to the peak of your profession as you are ever going to be.

The problem is that you are utterly out of shape, with no idea of how to get into it. I ask you to find something that suits you and start. Let us presume you have you have decided to swim. Don't dither – get into the pool and test yourself. Do what you can. At 50 you can realistically set your goal to swim a kilometre. You may not find this easy but it should not be too difficult. Get a book on swimming or do some research on the internet. Find out how to develop your swimming muscles when you are out of the pool. The exercise intensity should be sufficient to raise your heart rate to around 60–70 per cent of your maximum heart rate. Use the widely circulated formula of maximum heart rate (MHR) = 220 minus your age to reach a reasonable figure on which to base your exercise.

By the time you manage that kilometre you will begin to realise that you have improved a great deal. Remember, we are not concerned with time here. You may not be a natural swimmer. You may have short arms and stumpy legs but you like moving through the water. Great! Go at your own pace. Try doing it at least three times a week. You are only 50, there are decades ahead of you so do not set your goals too low. Keep testing yourself. Play a game with yourself, setting new horizons as you reach old ones.

If you feel that you are now engaging in an activity at an adequate level for the lifestyle you wish for yourself then there is no need to push further.

The Lazarus Strategy

At the age of 60 there is still time to do great things. My advice is very similar to that which I gave to the 50-year-old. I am assuming you have not yet deteriorated to the point where you require medication just to function. If you do, get clearance from your doctor first. Perhaps you could take up ballroom dancing, or hill-walking if there are hills nearby. The world is your oyster. An exercising 60-year-old could cycle around the world if that was their wish.

At 70, another decade has passed during which you were sedentary. Be more selective in the physical activity you choose at the start. Brisk walking is good or, for the more adventurous, cycle spinning classes can be fun and you can regulate your pace. Again, I know nothing about you, so advice specific for your needs is not possible.

Whatever you do, remember that while it must bring enjoyment, the activity should also push you and raise your heart rate to *at least 60 per cent of its maximum*. I suggest you worry about how much exercise later. Educate yourself. Get specific information on what you enjoy. All this does not require much intelligence. It merely requires diligence. How restrictive are your goals? Think on this: I heard a 77-year-old being interviewed on BBC radio because she had just completed a round-the-world solo sailing trip. Recently a well-known personality who is 54 years old ran 27 marathons in as many days for Sport Relief. The limits of your abilities are in your mind much more than they are in your body. Just go for it!

Oh, oh, here comes that ageing slope

As you age and your powers slowly diminish, you will need to adjust your goals. That is fine. Adjusting goals does not equate to not challenging yourself. Later on we will follow your performance profile as you age but for now accept that your capabilities are going to diminish, especially as you approach the 75-year mark. You will need to change the tempo of your activities every so often. Indeed, slightly more often when you enter your eighties. Believe me, I know.

Reading the above, you will also immediately realise that the time to start your new lifestyle is while you are still young, relatively speaking. Any time later will always be more disadvantageous as you will be starting with less potential ability. Instead of being among those of your peer group in tip-top condition, you will be playing catch-up. You can and will improve, but you have limited your goals. Please do not let that stop you. There is much evidence to show that any small increase in physical activity will significantly help you to achieve life-changing health improvements.

Attaining goals as an indicator of sufficient activity

Setting goals is a rough and ready way of judging where you are on your battle to healthy ageing. Let us presume that at 70 you have decided to considerably broaden your horizons and set yourself the target to cycle 50 kilometres. This is a great target because at your age it is well within your latent capabilities. Every week, make time to cycle and you will feel and see your improvement. Hopefully you feel good, you are not overeating and your muscles are doing what they were made for. The trinity is in command. Attaining your 50-kilometre goal means that you are doing sufficient physical activity. Do not underestimate yourself and set your goals too low. As a newcomer to physical activity you may at first be sceptical of your capabilities, but after a few months your prejudices and horizons will quickly change.

Do you wish to expand your horizons even further? Fine. Increasing the intensity and amount of exercise will not make you more healthy but it will make you more adept at your chosen discipline and so allow you to compete if you wish. You are now entering the region that competitive people inhabit. If that suits your personality then go for it but remember that the risks of injury increase the more you stress your body. These can vary from feeling fatigue after too much exercise to sore joints, shin splints and pulled muscles. If you do decide to push your own particular discipline then make sure you know the warning signs that come with the extra stress.

The crucial question is: do you truly enjoy your activity? The fact that you are enjoying yourself means that your mind, your physical activity and probably your weight are in harmony. A happy mind will push you into trying new adventures, secure in the knowledge that the body has all the necessary qualities to fulfil your expectations. Also, this new-found energy will spill over into all those everyday activities that we continually come across. You will find that you can now climb stairs more easily, mow the lawn with enthusiasm (maybe), wash the car, play with the grandkids, hurry for an appointment or easily carry your shopping. All these everyday activities are reliable indicators of your physiological health.

How often should I do what I enjoy?

Again, the answer to this question depends on the goals you have set for yourself. Remember, just doing exercise because you believe that is the way to change your life is mindless. Have you set your aims in advance? If your first walk around the block took you 20 minutes, that was your baseline. If you wish to decrease your time, then you may have to do your walking more often or perhaps push yourself harder. If you are enjoying what you are doing you may well decide to increase the time you spend doing it. I leave it to you to decide how often you will attempt to reach your goal. For sure, you will need to do the exercise more than twice a week.

Will you stop once you reach that goal? That again is up to you. But remember you are now standing on the brink of your own distinct physical greatness. You will be astonished at your latent capabilities. If it suits your personality, go for it.

Resistance exercise

Not all physical activity has the ability to strengthen bones or increase muscle strength. For these effects, resistance exercises are a necessity. Although both men and women should do these it is especially important for older women who need to work on their bone density given the changes in the body post menopause, and to ensure that they can maintain their balance and prevent falls. So, while there is considerable latitude in choosing a physical activity that you enjoy, we all must incorporate weightlifting into a healthy ageing strategy. For many people, this conjures up a picture of hulking bodybuilders in a gym. This linkage is incorrect. We do resistance training not to build great big muscles but to ensure that we *maintain* muscle strength appropriate to our age.

You can do resistance exercises at home or with your friends. There is no need to enter the testosterone-laden atmosphere of a gym. Use weights which will not push muscles to their absolute limit but will load them to about 65–70 per cent of that. How do you determine this level? You should be able to perform between 8 and 10 repetitions of an exercise with a suitable weight, but no more than that. If you can easily keep going after 10 reps, the weight is too light. If you can't even do 8 repetitions in good form, the weight is too heavy. Instead of weights, you could use kettlebells or resistance bands. A very useful guide on resistance training is put out by the easily acquired American College of Sports Medicine called *Guidelines for Exercise Testing and Prescription* (any recent edition). While this book is directed towards study, it includes many nuggets of information about exercise. There are tables showing VO_2 values for different ages in men and women, as well as information on various exercises. The first half of the book probably has much information that will be useful and interesting to you.

Reading the resistance training literature, you might get the impression that we evolved by following the decimal system.

Putting Activity into Practice

Everything in the gym tends to be done in three sets of 10 repetitions. The Sumerians followed a sexagesimal system, a system based on sixty. This is a very convenient system because sixty can be evenly divided into many fractions: sixty minutes in one hour, sixty seconds in one minute, and so on. The Mayans had a system based on twenty.

If you do not like ten, try a system that suits you. How about two sets of fifteen repetitions? I might just go for four sets of seven and a half as soon as I work out a method. Create your own goals. Do what suits you to attain them. Go with the flow or find your own. There is no absolutely defined road. And be careful when you consult YouTube. Those young, testosterone-fuelled, self-absorbed workout fanatics may be a bit over the top.

About 76,000 people – mainly women – fracture their hips every year as a result of falling. Lack of muscular strength and thin bones are central to the causes of falls (Healthcare Quality Improvement Partnership 2018). Resistance exercises increase muscle strength and strengthen bones. If there were an alternative to resistance training to keep those muscles up to scratch, I would certainly pass it on. However, there is none. If anybody tells you that non-resistance exercise is just as effective in keeping up or increasing skeletal muscle strength, do not believe them until they have proven it to your satisfaction.

How do you know when to stop increasing the weights? Do this when your physicality matches your lifestyle and when you can comfortably realise all those aims you set for yourself. Once you are able to comfortably do all the things you wish you had the strength to do, you've found the sweet spot. Remember, age will slowly drive the weights that you are working with downwards. That is life. Get used to adjusting your lifestyle through the decades.

Get away from mindless physical activity

Just remember: do not haphazardly adopt a regime. Mindless physical activity is a soul-destroyer. Do physical activity with a purpose. Get those 'alive' feelings coursing through your body. Whatever your choice, find out which muscle groups need to be built up by resistance training. Remember that resistance training must be incorporated into whatever discipline you choose. Take charge of your body.

Don't view physical activity as a weight-loss agent

Physical activity is wrongly viewed as a great way to lose weight, which often results in inappropriate behaviour. There you are, sitting down to a Christmas meal which, for some reason that now escapes me, requires all attendees to overeat. Some of those sitting around the table will say to themselves, 'I'll go for a walk around the block to offset all this overeating.' To coin a saying: fat chance! You would have to walk around the block all day to make any inroads into the surplus calories obtained from your excessive eating.

Someone else at that table will make a resolution that when the holiday season is over, they will definitely enrol in a gym.

In the New Year they do exactly that. They then engage in energetic physical activity for a month or two and find that this activity has done absolutely nothing to lower their weight so they give up.

This message cannot be any clearer. Physical activity is to improve the health of your heart and lungs (and everything else), tone the muscles, make you feel good and keep your mind on its toes, ensuring that it keeps supplying new ideas on how to use your new-found capabilities for contentment and happiness. That is the function of physical activity par excellence. Let us follow up this unfamiliar idea about the poor relationship between weight loss and activity with some hard facts.

Physical activity is not the way to lose weight for most of us

In 1997, twenty-five years of accumulated research on weight loss was examined by scientists (Miller et al 1997). It seems that in most of the studies examined, subjects were around forty years of age and just moderately obese. After a fifteen-week period of diet and/or exercise, the diet-alone subjects lost about 11 kilograms. With exercise alone, participants lost only 4 kilograms.

What about with diet and exercise together? Those lost 11 kilograms – the same as diet alone. Yes, you read that correctly. Those twenty-five years of study had found that exercise, in ordinary people, had little effect on weight loss.

Interestingly, no older people were studied. We rarely are. This deficiency keeps popping up in research into health while ageing.

The Lazarus Strategy

Another large analysis was published in 2012 (Thompson et al 2012). One study in the paper analysed involved a diet with a 25 per cent calorie reduction for one group and a 25 per cent restriction plus exercise in the other group. The study lasted six months and there was no difference in weight loss between the two groups. The authors do remark that exercise may produce changes in fatty tissue that calorie restriction does not. That is fine and provides further information that physical activity should be used not primarily for weight loss but to improve and hone all systems of the body.

A third study examined the effect of diet and exercise on weight loss in women (Kraemer et al 1997). Thirty-one over-weight women were put on one of four regimes: a control group, a group with diet-only restriction, a group with dietary restriction and aerobic resistance exercise, and a fourth group with strength training added to the diet and aerobic exercise. After twelve weeks the three dietary groups demonstrated a significant loss in body mass. However, no differences were observed in the magnitude of loss among groups. Those who trained increased maximum oxygen consumption. From our perspective, this study once again enforces our knowledge that exercise is a body toner rather than an instrument for weight loss.

Let us look at a fourth study (Wu et al 2009). This examined the results of 18 studies and found that the difference in weight loss between diet alone and diet plus exercise, on a two-year follow-up, averaged 1.64 kilogrammes. Peanuts!

You may need to lose weight, and you may also need to get more exercise, but consider these as two separate (though related) matters. Physical activity is to simulate your evolutionary heritage which, as we've observed, is linked to your ability to hunt. Its purpose in modern times is to keep us moving towards health.

Perhaps I can add a little nugget of information to your thinking. As you age, your ability to do physical activity will decline. That is without doubt. What is also sure is that the number of calories in a kilogram of fat remains exactly the same whatever your age. You immediately twig that in your later decades weight control is going to depend heavily on eating and not activity. I know. Why not get into the habit of controlling your food intake now rather than wait for your decline?

Putting the role of physical activity into an evolutionary context

Let's go back to the lives of our forebears. The hominoids are sitting around doing very little. They have not eaten for a few days. Tummy signals to the brain and says, 'I'm hungry.' Brain thinks about this and decides to hunt. For this, physical fitness is required. Hunting is not easy. You chase the prey, catch the prey, kill the prey, carry it home and eat it. Hunger goes away. You sit around again doing very little.

Notice the progression. From tummy to mind to muscles to food. You can immediately grasp that this evolutionary pathway has great survival value. It is basic for life and for survival of the species. Physical activity as a mechanism used specifically to reduce the body's stored fat was probably never a necessity. These hominoids were not overweight.

In modern times, however, the body's stored fat is often considerable, as we've discovered. This fat can supply energy for months. Our tummy–brain–muscle connection is all screwed up. Our brains no longer ask or require our muscles to acquire food. A totally different reason for activity is now brought into play. We expect muscles to work, not for hunting, but as consumers of stored energy. We are asking 'machines' – our bodies – which are geared to use energy as efficiently as biologically possible, to now 'remove' energy that is stored in the highest energy source available, namely fat. There is a major disconnect between the amount of energy used when most of us exercise and the huge depots of energy that need to be mobilised. Somehow, I think this use of muscle was not integral to the survival of early hominoids.

A kilogram of fat contains 7,700 calories. The amount of exercise necessary to burn significant amounts of stored fat would be huge. Let's say we need to burn 3,000–3,500 calories a week through physical activity, to offset and make inroads into our increased and stored food intake. The average young person uses about 300 calories an hour when strenuously exercising, so this would amount to 10 hours of intense exercise a week. Where are the vast majority of people going to find this time? And how many hours would an *unfit* 75-year-old need to burn off those 3,000 calories in a week?

The levels of physical activity might be easily reached by professionals but these levels are way beyond most of us oldies. The most I would be willing to concede about the use of physical activity for weight loss in us oldies is that it may play some part but that the effect is very much like having my grandchild help with the washing-up. There is an enthusiastic expenditure of energy focused more on taking part than on measuring outcome. I am going to leave it there.

Physical activity in the hospital setting

Although this book does not deal with medical patients, I would like to take this opportunity to describe just how effective physical activity can be as a therapeutic agent. Recent research by a group in Pamplona – yes, where the bulls run – shows impressive results (Martinez-Velilla et al 2019).

They studied a group of hospital patients with a mean age of 87 years who had been admitted with conditions requiring eight days of acute care. During their stay, the experimental group received five sessions of exercise per day. A similar hospitalised group did no exercise at all.

What did researchers learn when they compared the two groups? It seems that being in hospital and lying in bed doing nothing, even while under medical care, is not good for the elderly. The group who were hospitalised and also exercised scored better on a wide range of tests including cognitive and mood status, quality of life, handgrip strength, incidents of delirium, length of stay, re-admission rates and mortality. And this positive response persisted three months after discharge. This is a formidable list of improvements, way beyond the reach of any current pharmaceutical treatment. And all after just five days of exercise.

An exercise regime for patients in their mid-eighties

I'm sure you would be interested to learn what physical activity regime those elderly hospitalised patients were asked to perform. They were requested to walk at their own pace for 20 minutes, get up and down from a chair for three sets of 12 reps, twist a towel as hard as they could manage, and climb up a staircase of 20 steps (Isquierdo et al 2016). These rather easy basic tests were designed to test the strength of people who were elderly, non-exercising and in hospital.

I detail that programme to make a point. If this short bout of physical activity can make such a big difference to elderly patients, think what advantages you could gain by embarking now on physical activity while you are still relatively young and healthy. Read the biographies of the 80-year-olds later in this book for further insights into the advantage of taking steps to improve your health.

Never stop being active

It was late February and I went for a cycle ride. It was cold but I was well wrapped. I had not been out for a few months, and my legs were complaining. I managed 40 kilometres, cycling like a tortoise on Prozac. But I made it home. That is not too bad for 84. When the sun comes out, I will try again. If I can go further, that will be great. If I cannot, still great. I am quite aware that at my age I am like a downhill skier, accelerating as

Putting Activity into Practice

I descend. Join my wife and me, like the defenders of Masada, or the fighters at the Alamo, or a gladiator in the Coliseum: we know we are going to die but we will go down fighting. What is the alternative? Loss of independence, reliance on the state, doctors, medications and meals on wheels.

CHAPTER EIGHT

Putting Brakes on Your Eating

Is obesity a disease? A thought-provoking question. Americans have fairly recently classified it as such. Reading between the lines, the driving force for this view is that it allows doctors there to charge fees when overweight patients attend their offices. Labelling obesity as an illness, though, tends to remove any responsibility from the 'sufferer'. Label obesity with any term you wish. Call it what you will. I do not care. In order to age healthily, obesity must be tackled head on. There is no other option. It's strange how, during periods when food is scarce, obesity decreases. Wartime Britain was a good example.

Failure of messages about eating

This saga of obesity and its control is a story of abysmal failure of gigantic proportions. A 2004 article in *The British Journal of Sports Medicine* says 'traditional treatment strategies and public health interventions aimed at reducing the incidence of obesity are proving inadequate

at controlling the global epidemic of this condition'
(Davey 2004).

Because so many people are now overweight, we can agree
that dietary messages, like those about physical activity, are
not getting through. We have seen that exercising more and
eating less both require alterations in our behaviour. I do
think that messages around eating and food, like those about
activity, are not doing that for which they are intended. How
many times in this book have we encountered this problem
with messages? Many!

The ramifications of inadequate education permeate
throughout society. It is easy to give correct advice to an audience of one. It is an entirely different undertaking when trying
to give this message to a spectrum of people who have thousands of different approaches to food. When working at the
Department of Health, it was necessary from time to time to
provide recommendations on some aspect of eating and I had
extraordinary difficulty drafting something that would appeal
to all the public. In this book I am putting the onus on you.
You are the only person who can devise a message that applies
100 per cent to you.

Looking at yourself

If you are overweight, then reducing weight is an excellent place
to start your journey to better health. The idea that you can be
fat and fit is a myth. It is probably based on the equally false
idea we discussed already: if you are not ill, you are just fine.

Why do we gain weight?

In general, we have seen that activity can't save you from obesity, that's true. Being overweight is the baggage of simply eating too much. Sitting in front of the television and eating, or going to the cinema and consuming popcorn or ice cream, or watching a play and feeling we need sweets during the interval, or snacking on the way home from work and then enjoying a big dinner – and believing that a stroll around the block will make up for it. This is all a recipe for disaster. Where do we get the idea that sitting in a seat for a few hours consumes so many calories that we need to stock up again before we starve? This same question applies to eating when we are bored, or when we need comforting, or when we do not feel like it but we eat because everyone else is (like on a plane or ship, for instance). Why keep a snack in your handbag or your car just in case, despite your stored calories, you are overcome by hunger before your next meal. On top of all that is the temptation to always overestimate the energy burned during our last workout (or the one upcoming). It all adds up, like a snowball gathering snow as it rolls.

Food, glorious food

We live in an age where food messaging is always within 20 metres of us. Food is in newspapers, on television, the internet, on the pavements and magazines. There are shelves upon shelves in bookstores devoted to it. What to eat. What not to eat. Cooking programmes star chefs who prepare dishes totally inaccessible to us home cooks. Not only inaccessible but at times downright unhealthy. In any event, food is all around us in ways that it never was before. The pervasive idea behind all this is that we need to eat constantly; otherwise we will be danger of starvation. Most food today is designed not for nutritional value but to entice us to spend money.

Here is what I do. I never read any article on food. I never watch cooking programmes on television and I do not give a jot about some 'magical' new food ingredient that will transform my life. I do not care what the latest press release has to offer. The basics for a healthy diet were laid down a very long time ago.

The way I see it

The usual approach of providing prescriptive eating plans in order to get people to lose weight has been a dismal failure. How much of a failure? A 2019 report states that 29 per cent of adults are now classed as obese. This represents a 26 per cent increase from 2016 (House of Commons Library 2019). Another report states there were 711,000 hospital admissions where obesity was the primary or secondary diagnosis. This represents an increase of 15 per cent over 2016/2017. Around two out of every three patients were female (NHS digital 2019). It seems to me that the population is becoming familiar with these horrible statistics, and with familiarity comes an atmosphere of completely misplaced equanimity. Why should I worry when everyone looks like me? Oh my! Oh my! All understanding of the interaction and necessity for the trinity has been lost. A collective self-inflicted amnesia permeates our nation.

Different food for different people

We all have different likes and dislikes. The food we eat differs from person to person, household to household. There are as many different diets and ways of food preparation as there are nations. Look around your town or city. The vast panoply of human differences continually passes before your gaze. Most people keep to their own specific dietary favourites and culinary idiosyncrasies. You cannot expect me to offer some sort of catch-all diet that all people will want to follow. It would smack of arrogance and would certainly be the road to failure. Nearly all the nutritional advice given in the popular press requires us to change the foods we are used to eating but the antecedents of our current diet probably trace way back to our parents, grandparents and beyond that. Somewhere in your food lurk the ingredients of a healthy nutritional lifestyle. You only need to get rid of the junk and hold on to the good parts.

To my mind, the most rational advice on food and all its complexity was published by the Food Climate Research Network in April 2014 (Garnett 2014). This very interesting paper deals with how to keep a balanced diet, plus how to make diets more sustainable with suggestions on eating foods that may lower greenhouse gas production.

Those books on diet again

One almost totally useless way to lose weight is to buy a book devoted to weight loss. There are hundreds from which to choose. All have different recipes. All promise results. Every few months a new guru pops up who claims to have the universal secret for weight loss. Obviously, this secret has come to the author like manna from heaven, and probably not from contributing knowledge to peer-reviewed scientific research. Most books propagate a message of simplicity in the weight-loss process. This too is completely incorrect. To lose weight not only requires inner drive and the resolution to either switch to a different eating regime or to modify your traditional diet, but also to adhere to it for a lifetime. There is no easier way. Keeping to your correct intake is not just for Thanksgiving, Christmas, or for the New Year or any other holiday, but forever. Surely that does not surprise you?

Perhaps if these diet books also provided follow-up figures on how many of the people who bought the book actually lost weight, we might get some idea of the farcical nature of the whole enterprise. I understand that around 95 per cent of people who try to lose weight and keep it off fail. In a study published in 2008, the authors stated that the results suggest that strategies to increase adherence to a diet deserved more emphasis than the specific nutritional components of the diet itself (Alhassan 2008). I agree. The safest way out of this morass of half-truths is to make a big pile of these books and turn it into a home for your guinea pig or rabbit. Remember, they are both mammals, so if you build a book-hutch make sure that there is plenty of room for sufficient exercise.

Where do you start?

You start with yourself. Why does your profile resemble that of a pear? Asked another way, why are you fat? I am trying to get you to address the thoughts that are coursing through that brain of yours, the ones that drive you to do the things you do.

I was fat. I got fat because I drifted into eating more. I liked food served in well-rated restaurants, as well as consuming on a grand scale malt loaf, Bath buns, fruit cake and so on at home. I was well over my weight limit and still gaining at speed. When I finally awoke, like Rip van Winkle, I immediately decided to stop the drift.

Tentative first steps

If you are an overweight 50-, 60- or 70-year-old with much experience of life, then be sufficiently brave to choose a different path, your own path. In order to do that you need to begin with some idea of how much and what you are eating. That is your first goal. No half measures.

A clear and unambiguous way to start is to cease snacking. This is a no-compromise action. Lose the idea that you are like a browsing deer and need to nibble every few feet as you move. Yes, lose all those foods on which you have wrongly lavished unreturned love. I mean all crisps (whatever their origin), biscuits, nearly all the plastic-coated meals you put in your microwave (read the ingredients list), cakes, buns, confectionery and so on.

It is very difficult to keep track of how much you are eating if you never stop, and the tendency is to minimise or conveniently forget how many calories you have consumed. Keep to three meals a day. Keep a diary if it suits you. Remember Pavlov's dog? The fewer times you can get into an eating loop, the better.

Fighting those constant whispers to keep eating

Another problem when cutting down on eating is that it becomes very difficult to think of anything *except* eating. There is a constant gnawing noise in the background. In your head sit bad cop and good cop. Bad cop is reminding you that you must eat because you want to, good cop is telling you that this behaviour has abysmal consequences. Bad cop whispers, 'Do you *really* want to postpone instead of going for instant gratification?' It takes at least a month to bring this constant food craving under control. But it can be done.

In addition, eating is an important medium through which we interact with family and friends. Eating together means a lot to us. You will have to develop the fortitude to stand against this continual social bombardment. Now, you are beginning to get an idea of the resolve necessary to do what is an absolute necessity for your future health.

One way of many

If you still have no clear idea of how to start, let me suggest the following to you. Keep to what you normally eat but resolve to eat about three-quarters of your normal intake to start with. So instead of four potatoes try three. Get a set of scales and keep it near and convenient for use. Get a book that lists the calorie value of foods. I again recommend McCance and Widdowson's *The Composition of Foods*. Put out by the Royal Society of Chemistry and the now defunct Ministry of Agriculture, Fisheries and Food (MAFF), this lovely book lists just about every food that can be purchased in a supermarket and gives a breakdown of every food stuff, ranging from major food groups to inorganic components such as minerals, as well as vitamins and everything else necessary to be properly informed about your food. You can also use www.nutracheck.co.uk.

Weigh the potatoes and get an idea of how many calories there are in the helping. Be scrupulous. Do not cheat. Weigh all the main ingredients of your meal, the things like meat, fish, grains, rice or any other food stuff that makes up a good portion of your diet. Work out your calorie intake.

Foods like green vegetables can generally be ignored, as can most salad ingredients, but beware! The addition of huge amounts of mayonnaise, oils or fried croutons to green salads immediately change a salad from a very acceptable food to one that can be high in calories.

Read your book on the calorie and nutritional value of food. Become your own expert on what to avoid. If you can hold out and keep to your programme for a month, there is real hope. You have probably beaten bad cop by then and also got your friends used to your new eating habits. Your ability to abstain will tell you whether you are ready to progress. There is no hiding place. You are either with it or you are not.

I found that keeping a food diary was good because the action of making my entries made me pause and ask whether I really needed what I was going to stuff into my mouth. I found this behaviour kept me in check when visiting restaurants and I would inevitably choose the lighter options from the menus. This sort of discipline suits my temperament. You may have other personal stratagems that work for you. If you keep to the foods with which you are comfortable, you are not changing the landscape of your food, only fencing it in. Look at yourself in the mirror; you have enough stored food to last weeks. Nothing bad is going to happen to you, except your resolve is going to be tested.

Forget about weighing yourself. The results will tend to squash your optimism. Our weight varies from day to day because of water and other factors. Watch your waist instead, or just the way your clothes fit. Get out that pair of trousers or that dress that does not fit and make it your aim to reclaim your wardrobe.

After the month you should have the confidence to refine your diet and make it your own. You can begin to devote time to getting your calorie intake to allow you to attain a weight that is in keeping with your age and gender.

Enjoy teaching yourself. Look on the whole enterprise as a journey of self-discovery. Are you the person you think you are or are you kidding yourself? All it takes is a little time. Take charge rather than be the passive recipient of advice. There is nothing about eating that is very complicated, once the bells and unnecessary whistles have been eliminated. You need about four of the aisles in the supermarket, mainly those where fresh fruit and vegetables and fish reside. Get to know them well – you are going to have to do this for the rest of your life.

There is another bit of not-so-good news. As you age, your calorie requirement goes down. You will have to adjust your intake to this sliding scale. Do not get overly concerned, you are now developing the means to regulate your intake as you age. Remember you wish to lose weight not primarily for vanity, but to free your naturally endowed physiological gifts. Keep your mind, activity and food under check and looking and feeling good will follow. The utterly surprising finding will be that your new discovered self will pull you to wonderful adventures. If you have the courage and the will then the payback will be absolutely great and persist into old age, I promise.

The complexities of giving individual advice

Occasionally, friends ask me for advice. You will already gather that I do not relish doing this. It requires lots of time devoted to someone with one problem particular to themselves. I do not think that diet books address this problem of individuality and age at all.

I recall one neighbouring couple who approached me. They had previously been trying to follow the advice they found in newspapers, magazines, books and television, and had become totally confused as to what and how much they should be eating. I started by asking them to keep a three-week food diary. Every morsel had to be logged. On analysis it became clear that their diet had few of the correct ingredients for health. They had in fact become too scared to eat properly. They were so mixed up and frightened by advice that what they were eating was a mess. The new start began with gathering up all the books and articles that they had accumulated and turning them into mulch. I do not like burning books no matter how useless.

Next, I gave them talks on what to eat in order to empower them to take control of their lives. I let them decide how to integrate the advice into their new eating habits. They never read another article or bought another book on food. They ate what they liked in moderation, making sure that the mix of food they ate made up a balanced diet. They monitored their weight by the comfort of their clothes. They did physical activities that suited their lifestyle. But they did not use these activities to control weight. They had their weight in control by eating the amount of food they needed for their age and gender.

Keep 'eating less' and 'physical activity' in two compartments

I consider it sufficiently important to warrant revisiting this point because the idea of physical activity for weight loss is ingrained into nearly everybody. Here are the rules for us average folk:

- Eat the correct amount to attain and maintain your designated weight.

- Exercise to live a healthy, active life. Don't confuse those two, or their relationship. Do not worry about how many calories you will burn when you exercise. If you are like most people the number of calories you are using while you are doing your activities is small beer compared to the amount of weight you need to lose. These calories are also small beer compared to the amount of calories you can eliminate from your diet. In any case you are going to replace most of these calories before your next bout of physical activity. To believe that you will exercise sufficiently to make up for all your dietary sins is futile. You cannot outrun a lousy diet.

How June and I regulate our eating

I do not know a best method. June and I, when we were overweight, followed different paths. My wife with only a few kilograms to lose took what was probably the most sensible route. She cut down on all helpings. She modified her portions as time went along. She shed weight slower than I did but the end result was the same. What weight did I aim for? I didn't – I simply decided I would get down to the shape I was in when I could wear 32-inch waist trousers. That was 34 years ago, and I have kept a steady weight since then.

Checking my diaries, I found that I opted for about 800 calories a day. I opted for a Very Low Calorie Diet (VLCD). There is a large body of literature on this subject. Read all about this regime before embarking on it (Tsai and Wadden

2006). You may need medical clearance. Ten weeks later I was at my goal, and my old trousers welcomed me back. I then began to slowly increase my diet to over 1,000 calories and then to about 1,800–2,000 calories, all the while keeping an eye on my trouser fit. Now I have a good idea of how much I am eating. I eat everything that I like but in moderation. I eat three times a day. I do not eat between these meal times. That way I do not have to face or think about food every few minutes. I think that the more often you eat, the more you risk becoming Pavlov's dog.

I believe absolutely in 'calories in must equal calories out'. All the rest to my way of thinking is bells and whistles. Of course, this way may not suit you. That is fine, chart your own course. I was lucky in having a like-minded companion as having company makes it easier. If any food police read this and the advice is not to their liking then please inform us why nearly all the other advice offered is an abysmal failure.

How we keep on keeping on

June and I do not eat already cooked food that comes wrapped in plastic. We have the necessary 40 to 50 minutes to prepare meals from scratch. Why would you wish to eat all that stuff that is listed on the label? Over our 80 years of living and our 55 years together, we have refined our food to that which we have come to enjoy. The list is not the result of being afraid to try new foods. It's the result of having tried a very large variety throughout our lives and coming to an understanding of what suits us. June and I require only those four aisles of the supermarket to satisfy our needs most of the time. We eat plenty of vegetables. This regime relegates food to a fixed time and helps prevent the Pavlov's dog routine. Pouring a cup of tea or coffee or a small piece of chocolate or nibbling a cookie initiates the flow of gastric juices. You then feel hungry and start the whole eating process.

For us, breakfast is porridge or muesli, fresh fruit and something to drink. Lunch is very light with either a salad or sandwich and sometimes a cup of tea. We are careful with portion sizes and eat very little animal protein. The average human, whatever that means, needs only about 100 grams of protein a day. This can come from any source – animal or vegetable.

We prepare sufficient food for our needs and we do not make too much. That is another sneaky way of encouraging yourself to eat more.

Many restaurants serve food somewhere between awful and mediocre. That is why guides are a necessity. For this reason, June and I have refined our choices of which restaurants we'll frequent and which foods to choose. Cooking at home beats over 90 per cent of restaurants.

How as a society are we going to control overeating?

The NHS is being overwhelmed. The *Guardian* published a piece which claimed that two-fifths of the NHS budget was being spent on those over 65 (*Guardian* 2016). An 85-year-old costs the service seven times more, on average, than a man in his late thirties. To put it another way, about a fifth of the population is consuming twice as many resources compared to the rest of the population, and the ageing population is set to keep rising. Do you think we should change our behaviour by our own free will or would you like the government to step in and ban all sorts of habits, and insist that we do physical activity or pay more tax? I do not know what all the options are but you get the general idea. I am betting my cotton socks that unless there is a self-motivated, radical change in how we eat, we oldies are sooner or later going to face a rationing of the health service.

CHAPTER NINE

The Set Point Theory

As we have said already, exercise is not a one-size-fits-all proposition. Because we are all different, the amounts of physical activity we need will vary from one person to the next. It will also change depending on our age and other circumstances.

Let us examine the effects of exercise on, for example, the heart and get some idea how that particular organ reacts to exercise.

In 2014, B.D. Levine and his group, who have extensively studied the heart, published an interesting paper in the *Journal of the American College of Cardiology* on the effect of exercise on heart function: 'Impact of lifelong exercise "dose" on left ventricular compliance and distensibility' (Bhella et al 2014). In plain English, they were studying the efficiency of the heart in pumping blood after having been exposed to various levels of exercise.

We can regard the experiment as investigating the effects of different doses of exercise. 'Dose response' is often measured by drug researchers in order to ascertain the effective therapeutic dosage of some drug. They are looking for the lowest dose of medicine that causes the desired therapeutic effect. In this heart experiment, four different 'doses' of exercise were investigated: (a) people who were sedentary and got no exercise at all, (b) those who did a little bit now and again, (c) committed exercisers like myself, and (d) competitors, who were exercising about 6 to 7 times a week. This is well above what most of us do. There were about 25 people in each group, all were around 70 years old.

The results

Group A, because they were sedentary, had the least efficient hearts. In Chapter 5 we looked at the physical capabilities of sedentary people, and from our discussions would place Group A firmly in the category of *not ill but not healthy*. The performance of their hearts re-emphasises this categorisation.

Group B showed that exercising haphazardly had no advantage over being sedentary. It reinforces the idea that we must take exercise seriously. Strolling round the block, going shopping and pruning the roses is just not good enough. Remember the 60 per cent of MHR rule.

Group C are people like me. We are exercising sufficiently to be well above the danger zone shown by groups A and B – exercising at a level that allows us to be free of those 26 diseases of ageing that we have previously met. In short, we have a healthy physiology.

Group D are those who have decided they wish to compete and accordingly increased their exercise levels. Notice they are now way beyond the exercise levels necessary to keep diseases at bay (Group C). They have the necessary enhanced physiology to be competitive but they are not healthier. As I have stressed time and again, and I will revisit shortly, there is no evidence that competition-level physical activity produces superior health compared to those who manage regular strenuous exercise. It produces superior athletic performance but that is another whole different kettle of fish.

It is worthwhile pointing out that our old friend VO_2 Max also has been shown to have a dose response to exercise and, not surprisingly, the sedentary do not do well.

From the above data we can now begin to think in a different way about how much exercise an individual requires. Clearly there is an amount of exercise which moves one beyond the horrible risks associated with people who mainly sit and watch television in their free time. Let us call this level of exercise the 'set point' (Lazarus and Harridge 2017).

How do we determine our personal set point? There is not a straightforward answer. Everyone differs so set points will also differ. What follows is my way of using set point.

Those of you who have been exercising for a considerable period will have an instinctive feeling for the state of your body. You need sufficient physical activity to enable you to comfortably do the everyday things that all humans should be able to do. If you are middle-aged, can you run or walk rapidly for 25 or 30 metres? Can you run for a train or bus without getting breathless? Can you climb three or four flights of stairs without puffing like a steam engine? Can you cycle 5 miles comfortably? Can you swim half a kilometre? Can you easily carry a sleeping six-year-old child from the car to their bed on the first or second floor? If the answer to these questions is no, it's a safe bet that your activity level is not meeting your set point. And so you must do more. We all have to adjust our goals as we grow older. But we must maintain basic physical capabilities at a level that allows us to live independently and healthily.

You will see that trying for a set point harks back to the advice given when you first contemplated taking up and increasing your activity levels. Concentrate on setting challenging goals for your age. Pursue those goals by the means you have decided is right for you. When you can do what you set out to do you are probably sufficiently near your set point to be confident that you have banished the effects of a sedentary lifestyle. Do not get hung up on the concept of set point. Think of it as a way of conceptualising the relationship between you and physical activity. Use it as a mental aid to keep you doing sufficient physical activity to offset the problems associated with too little exercise.

What happens if you keep increasing physical activity beyond set point? You will find that you increase your ability to tackle harder goals. You may also be moving into that area where, if you so wish, you begin to equal the physical prowess of competitive athletes. We might call this the master athlete class. Hurray for you. But there is as yet no reliable evidence that master athletes live longer or have a shorter period of morbidity than those who exercise to their set point and no further.

You can now easily answer the next question. What happens if you exercise below set point? Obviously that is insufficient activity and you fall into the A or B groups described in the heart experiment. There is no difference between what you are doing and doing absolutely nothing. Get out of the danger zone. Get out and move.

CHAPTER TEN

More Facts on How Exercise Affects Ageing

We all operate under the belief that proper diet, activity and a sound mind – our friends the trinity – will help keep us youthful and vibrant and healthy; and, yes, this is a comforting thought and it is true.

I wish to relate some important new findings from research conducted by my colleagues and myself (Lazarus, Lord and Harridge 2019). What we uncovered is that not all body functions or systems respond to physical activity in the same way. There are variations. Let us have a look at these types of responses and I will explain the way I view this new research.

Category A

This category contains those physiological and biochemical functions that are influenced solely by the ageing process – meaning that nothing we can do will improve them. Get plenty of exercise and eat the healthiest diet possible and it will make absolutely no difference where these are concerned.

One example is maximum heart rate (MHR). This is simply the average number of times your heart should beat per minute during vigorous exercising. This is determined, as you might expect, by having you engage in vigorous exercise and then measuring heartbeats per minute. But that method can be circumvented with the mathematical formula to predict your maximum heart rate (MHR = 220 minus your age). Figure out your own MHR if you so wish, mine is $220 - 84 = 136$. That is close enough. MHR bears some relationship to cardiac output but let us not get into detailed cardiac physiology here. It is not necessary.

What can we deduce from this formula? Notice it asks only for your age. Nothing else. Where MHR is concerned, nothing we do matters – the only important factor is our age. It is quite sobering to find that your heart has a mind of its own.

Another function in this category is verbal fluency and memory. These mental functions decline with age. Exercise also does not seem able to affect these functions. Don't we know it?

Category B

This category includes some of the processes that typically decline as we age but *can* be influenced, to a degree, by exercise. One example is our old friend VO_2 Max. We have already seen that this is an indicator of all-cause mortality. It naturally decreases as we grow older. However, we can raise VO_2 Max by physical activity, especially by doing endurance exercises like running, swimming and cycling. Obviously, it will still fall with age but we have given ourselves a tremendous advantage by pushing it up to a higher level.

Muscle function can also be maintained by exercise although it too declines, inevitably, as we grow older. It is good to know that we can affect muscles positively by exercise, and particularly by resistance exercises.

Category C

These processes do not change with age – which is a comforting thought. It might come as a surprise that not every system changes with age. However, here there is more good news. Exercise will improve their function. An example is resting heart rate. If you engage in regular cardiovascular exercise, your resting pulse will slow down. What is the advantage of a slower heart rate? One advantage is obvious. Let us say you have a rate of 72 beats per minute. Simple arithmetic will tell you how many times your heart will beat in a year. This works out to about 38 million beats. Now, exercise regularly and your resting rate drops to 50 beats per minute (about 26 million a year). Obviously, over a lifetime you are saving your heart a lot of work.

Category D

Here reside the most surprising systems. These do not change with age. In addition, they are impervious to exercise. They go their own merry way through the decades. Again, it is rather comforting to know that there are processes in our body that refuse to follow the usual deterioration that occurs as we get older. An example here is the function of the small intestine. It just keeps working away, feeding us nutrients that are obtained from digested food, and it doesn't care how old we are.

Putting A, B, C and D together

Try to imagine how many different processes and functions are going on in your body; I cannot even begin to count. You will note that I gave only one or two examples in each of the four groups. Obviously, there are many, many more functions to consider and categorise. This research is still in its infancy, and most researchers have not yet started to classify systems in this new format.

Four Categories of Regulation While Ageing

Only B and C affected by Exercise

Category A	Category B	Category C	Category D
Purely age dependence	Age dependent but changed by physical activity	Age independent but changed by physical activity	Independent of age and physical activity
Maximum heart rate	VO$_2$ Max	Resting heart rate	Small bowel function

More Facts on How Exercise Affects Ageing

If we could peek into our bodies as we age, we would uncover a mosaic of regulation. Some processes going downhill with age, others being supported by exercise, and some unchanging no matter what we do. It is good to know that we can directly affect at least 50 per cent of what is going on in our bodies. It is also clear that by influencing those systems that can be influenced, we improve our health and optimise our ageing. The ageing process may not always be our friend but under certain conditions – meaning when we take care of our activity and eating – we will age as gracefully and healthily as is humanly possible.

CHAPTER ELEVEN

Of Mice, Us and Ageing

The effective treatment of disease was probably ushered in by the antibiotic era. Antibiotics were, at their beginning, always curative. Patients took them, got better and the disease went away. With the discovery of other different diseases came the necessity to develop other drugs and the need to diversify the training of medical practitioners in order to ensure high-class treatment. There were about 61 different specialties last time I looked. Every disease now has a specialist. The tendency to view the body in smaller and smaller segments appears irreversible. It is from this disease-imbued perspective that most health professionals view human ageing. This perspective is also enhanced by the non-human models used to study ageing.

Caged mice and their relation to ageing diseases

In order to increase knowledge about diseases and the way they progress, animal models were developed. By far, the majority of animals now used are rodents. These models have been one of the foundations underlying the attempt to understand all human disease. However, we are not interested in disease models.

We need to ask whether mice, rats and humans are all on the same evolutionary ageing pathway. A report by the European Molecular Biology Organization queries whether this assumption has validity and comes to the conclusion that it might not be correct (Demetrius 2005). Immediately this raises doubts about their suitability as models for healthy human ageing. It is worrying that their ageing and ours might not be following the same evolutionary path. Let us be clear, we are not querying the suitability of mice for the study of disease but we are beginning to have serious doubts about their usefulness when it comes to researching healthy human ageing.

Another important problem concerns the conditions in which the mice live. They are caged, the human equivalent of sitting in front of the television all day. We know what that means in health terms. Living in a cage is not their natural state. Give a mouse the opportunity to exercise and it will run kilometres. So, from all our previous discussions we would deduce that most caged mice are already compromised as regards their health. I definitely think that caged animals are not suitable models for healthy ageing humans. Why should mice be any different from you, me or your pet dog in their physical activity requirements?

Thirdly, it was found that if the usual diet of mice is decreased by 25 per cent, they live longer (Biological Effects of Dietary Restriction 1991). This effect is not seen in humans, re-emphasising the difference between rodents and humans. A study involving around 100 people and stretching over two years examined the effects of calorie restriction on longevity (Ravussin et al 2015). This was a heroic effort but it's hard to say much about human longevity after just two years. I had a look at the data and as far as I can conclude, the results showed if calorie intake was reduced by 12.5%, as they did, then everybody who ate less lost weight. The link to longevity is speculation. If you have been paying attention to what is going on around you, you will have noticed that over the past several decades human longevity has increased *despite* our atrocious lifestyles (Office for National Statistics 2018).

What stands out from all of the above is that caged inactive mammals should not in any circumstances be used as epitomising healthy ageing. All biochemical or physiological measurements that are made represent the status of a *not ill but not healthy* animal. A term we have encountered before. This is dangerous territory. Cotman and Engesser-Cesar report that animals that exercise show an increase in brain-derived neurotropic factor (BDNF), a molecule that increases neuronal survival, enhances learning and protects against cognitive decline (Cotman and Engesser-Cesar 2002). All data on ageing obtained from caged, non-active mice reflects a disease pathway similar to that found in non-active human beings. We wait and see whether exercise affects the same pathways in mice as in humans. Does a mouse plan for its old age? Worry about its grown offspring? Modify its behaviour to accommodate age? Is the mouse's mind affecting its health? Is the trinity of ageing an integral part of a rodent ageing process? What do you think?

Advancing a different perspective on ageing

Let us presume that I am an 80-something man who watches his diet and engages in physical activity, and I am tested for physiological function in a laboratory. Gasp! Surprise! My ability to function is found to be the same as that of a non-exercising 50-year-old man selected from the general public.

'You have the physiology of a 50-year-old!' all the medics, scientists and supporting staff exclaim.

'Excuse me,' I say. 'Your perspective on healthy human ageing is totally wrong. I, the 84-year-old who exercises and moderates his eating, am exactly where I should be – in an 84-year-old's body. But that unfortunate 50-year-old is in big trouble – his physiology is on the road to collapse. And we know this because he has already deteriorated to the level of an 84-year-old – three decades too soon!'

What do you imagine that poor man's fate will be when he is actually 84? Assuming that he's still among the living, and anticipating that he will not have improved his habits? He will be just the person I described earlier – infirm and ailing, but kept alive by doctors and pharmacists. This scenario returns us to that state of being *not ill but not healthy*. That 50-year-old may not yet have clinical symptoms but chances are that before long he will cross the invisible, unnoticeable threshold that separates healthy from sick.

CHAPTER TWELVE

A More Detailed Look at the Inherent Human Ageing Process

All experimental studies on humans, whatever they may entail, are carried out over very short time spans compared to the number of years we live. The observations can be compared to a snapshot of our lives. A photograph of you today won't really tell me much about how you'll look in ten or twenty years from now. We can only guess based on today's image.

That is exactly what is happening in nearly all ageing studies in humans. There are many reasons for this. How could I keep monitoring a single person over the course of decades, or a lifetime? The cost alone would be prohibitive and, in the end, how much would we learn from studying just one person? Would the scientists monitoring the experiments all be retired before it ended?

OK, there are challenges. But if we could do it, what would we see?

I can suggest a way to find out – let's look at people who are moving and eating in ways that should ensure good health and high function. And let's see them in action not just for a day or a week but over the course of their lifetimes.

The International Masters Games Association

Every year, all over the world, athletes of all ages take part in championship competitions. These are not only for the young in their prime – performers between 35 and 80 (and beyond) take part. By examining the performance of athletes at various ages, we may learn something important. Theoretically, we should be able to see an entire lifetime of athletic endeavour in a single flash.

Let us look at both male and female master swimmers. Are they all doing the same physical activity, and at the same distances? Yes. Are there competitors of every age group? Yes. Are these individuals at or near peak human capability, allowing for age? Yes again.

Now, let's chart how their performances change depending on age. We are going to judge these athletes in a 1,500-metre swim. This particular event has existed for many years, so there is a great deal of data available.

First, we'll plot the finishing times for all involved (see Figure 1 on page 119). Then we'll look at those lines on our graph. We see a curve. Similar curves have been published in a number of scientific journals. It is from one of those sources that I have obtained the swimming results (Donato et al 2003).

The effect of the inherent ageing process

What we see in that curve is that performance for the 1,500-metre swim goes smoothly downward as the same swimmers grow older. This declining curve shows no indication of a disease process at work – there's no sudden impairment of ability or power. The decreased performance as the competitors age is synchronous and integrated. This is a portrait of the inherent ageing process at work, and it is beautiful to behold. An attempt has been made to describe this curve with the mathematical formula: $[y = 1 - \exp((T - T_o)/tau)]$ (Baker and Tang 2010). However, do not get too hooked on this, as I think numbers do not adequately describe the ageing process.

What happens if we change the distance?

Now, it becomes really interesting. Let's examine the curve produced by short distance swimmers at 100 metres. We see the same downward path, which can be described by the same formula. Are you trying to tell me that the performances of a sprinter and an endurance swimmer follow the same decreasing curve as they age, even though these events seem significantly different in terms of their demands on the human body? You got it in one guess. Of course, that would happen! Why? Because we all are subject to the same ageing process. Remember the bread cutting analogy.

But hold on – I am not yet finished.

The effect of the inherent ageing process on all track disciplines

Let us change sport. When runners' times were examined, the performance curve for 100 metres followed the same path as for 10,000 metres (Lazarus and Harridge 2017). Baker and Tang published in 2010 that the curves generated by *all* running events, from 100 metres to the marathon, follow the same decreasing pathway as all the competitors age. So, another prediction fulfilled.

The ageing process is independent of activity – short sprint or long-distance endurance race, swimming or running. I am telling you that neither the body type of an athlete nor their performance times have any effect on the way the inherent ageing process affects performance. Stop and consider how everybody currently thinks about ageing, and try to square it with these findings.

A bit more thought

The first thing naysayers will say is that these competitors have champion genes which make them different from the rest of us. So, what are these genes? No one can say. If only athletes aged like this, then we would expect their superior genes to allow them to live longer and better than the rest of us mere mortals. There is absolutely no evidence for this. It is therefore reasonable to conclude that, thanks to our evolutionary heritage, *all* humans have ended up with the same ageing process.

A Look at the Human Ageing Process

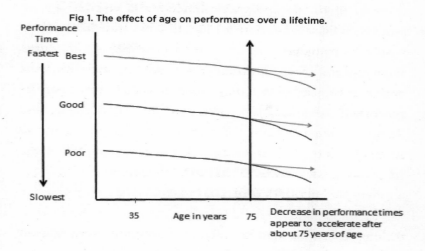

Fig 1. The effect of age on performance over a lifetime.

Poor swimmers and their performances

Let us try another experiment. What would happen if I decided to become a champion swimmer and then follow the decline of my performance through the decades of my life? Remember, I am not built for fast swimming but that does not matter.

Would the curve be the same? I am sure it would, because we healthy agers, providing something unforeseen does not come along, are all going to die round about the 90-year mark irrespective of our performance or build. People like me who have exercised at a fairly high level for 40 years or so are as healthy as they can be. I am sure if I took to swimming and had all the necessary training, my times for the 1,500 metres would be abysmal. However, because I am now a well-trained swimmer, and I eat correctly, the only factor driving the decrease in my performance would be the inherent ageing process, which is the same for all humans, athletic or otherwise. I am prepared to bet my bike that the shape of the curve showing decreasing performance would be the same for both poor and excellent sportspersons. If only there was a way to generate those data!

Interpretation of data from my perspective

We are back to the interpretation of data and the idea that we bring our own prejudices to the task. Provided all three components of the trinity are securely in place, the inherent ageing process is blind to the physical activity that we do. The ageing process operates in the same way for both sprinters and long-distance athletes, for the champions and for the plodders.

Incidentally, the shape of the curve applies equally to men and women. There appears to be no difference between the short, medium or tall when it comes to lifespan. Again, the inherent ageing process operates in a way which allows big and small physiologies to follow the exact same pathway. I have absolutely no idea how this physiological regulation takes place. No idea where the coordinating centre can be found, although I will wager that it sits in the brain. There is an enormous lacuna in our knowledge of human ageing. This lack of knowledge affects every person on the planet.

You and the performance curve

But wait! Let us try a final thought experiment. Do you mind if we use you as the model? You do not – thanks! We will designate you as a swimmer, a poor one, who has been training hard. You are the correct weight for age, sex and gender, and you are 50 years old.

Let's place you today at the top of the slope, indicating that this is your fastest swimming time (Poor curve, Figure 1). You are at your best. We now age you ten years. You have kept up your training. At 60, let us place you a bit further down the slope. As expected, your performance has gone down. This is due to the ageing process. Another decade goes by and you are still in training. At 70, your performance has further decreased, and we place you just at the point where the slope changes shape and begins to turn downward.

We know that the change in performance from decade to decade is due to the inherent ageing process. We are still functioning on all cylinders but all our physical capabilities and probably some mental capabilities are slower. Because you have all three parts of the trinity in place, the inherent ageing process is taking you smoothly downward. There is no hint of a disease here.

Go forward another five years. We now place you on that part of the curve that is beginning to slope even more steeply. I wish I could explain why this happens but I cannot. Thus far the news about ageing has been sort of reasonable in the sense that your decline in performance has been steady and even. After about 75 years the decrease in your performance simply picks up speed.

The Lazarus Strategy

Be philosophical about this. Just accept that as you approach the very late decades of your life your physiology is going to decline faster than it did when you were younger. Nobody can reverse time. Therefore, nobody can reverse age. Remember that please.

This ageing process that I know and the one which inhabits me, my friends, and you bears no resemblance to that monstrous 'disease' as depicted by experts who should know better. The ageing process, although moving us to an immutable end, may not be a joy forever but it is a thing of beauty.

CHAPTER THIRTEEN

The Lazarus Strategy in Daily Living

I thought my description of what I do on a daily basis might be of interest and might also provide a template of physical activity and movement that you may be able to use as you take your first steps into the world of movement and physical activity.

I am acutely aware that you and I are different. I am not being prescriptive but merely setting out some signposts. If you are older or less able, you will do less. If you are in your fifties then the sky is your only limit, if that is the limit you wish to set. No one can force you to do more than you wish, and especially not a book that is easily set down on a table and never picked up again.

The daily round

Eating

I need about 1,800 calories a day when functioning normally. I'm generally up and about by 7.30 a.m. Breakfast consists of 60 grams of jumbo oats porridge made with water. When cooked I add about 120 grams of grapes. Mix and eat. Why oats? Well, look at the ingredients list on a packet of oats. Nothing in the packet but oats. It is a naturally occurring complex carbohydrate food. To add sugar or cream would negate the reasons for eating this food. No, I do not weigh amounts out daily. After a week it is possible to judge amounts fairly accurately. Because of the added water I do not need to drink, but if I feel inclined I will have tea, either Russian caravan or Assam flowery. No milk. It spoils the flavour. (All in all, about 400 calories).

I do not eat between meals. That's it. When I attend meetings at the university, I may have a black tea, coffee or water but no food. I have taught myself not to assuage boredom by eating, or to overestimate the number of calories used sitting around a table. If you eat out of boredom you will become like Pavlov's dog, and boredom will signal your eating pathways to fire. I am always slightly surprised when after an hour of talking, people claim they are famished.

The next meal is lunch. Again, I do not eat much. I will have two slices of wholemeal bread with some cheese. (Any cheese will do. I like them all.) I will have an apple or another piece of fruit. (Adding up to around 400 calories.)

I do not eat bread that comes wrapped in plastic or is presliced. The last time I looked at an ingredients list on such bread, there were about 20 different components and chemicals. To my taste buds these ingredients make the bread taste like nothing that I care to eat. I suppose we must call it bread but it probably comes from outer space. I am totally unable to recognise the buns used for hot dogs or hamburgers as being related to bread as I know it.

That leaves about 1,000 calories for my dinner. June and I vary the evening meal. We sometimes eat meat-free. We have a slice of fish, 100 grams each, about twice a week, served with lots of vegetables. Nearly all of our vegetables are organically grown. We tend to vary our foods day by day. Remember that you only need about 100 grams of protein a day. Do not go mad on protein by eating a 500-gram steak, (the calories in that are probably sufficient for a day).

We will end the meal with fresh fruit or cooked plums served with fresh cream. As a good general rule, we try to avoid, as much as possible, supermarket meals that come wrapped in plastic. Again, read the ingredients list. Sure, packaged food is convenient but there are more components than a fresh-cooked meal would require. How come you are so busy in your seventies that you need convenience food? Could it be that because you are so sedentary, that you're too tired or weak to make something simple? We do not make a fetish out of food. We do not believe there is anything mystical about it. We eat out about once a fortnight.

Tonight, we just happen to be eating out. We are having Indian food with friends. I love plain naan bread baked in the tandoor. I will have lentils and aubergine. Probably chicken cooked in the tandoor oven. Luckily, I do not have a thing for sweets, although I do love apple or any other crumble. At 84 I have almost stopped drinking alcohol. I need to be clear-headed when driving and I find that even a 125 ml helping of wine can take the edge off my thinking.

My attitude to food

You may be tempted to remark that I have little appreciation of good food. When I retired, I decided the time was ripe to learn high-class cooking. So, I enrolled for a week's course at Raymond Blanc's school, which was held in Le Manoir aux Quat'Saisons. With the course came the opportunity for June and I to enjoy one of the lovely ground-floor suites. There were eight of us taking the course. The pupils were tutored to cook their own lunches and dinner was in the dining room. What a week. I learned to make a chocolate coffee cup and fill it with what looked like froth. It was totally useless for home use. Just too much work for too little reward. So, you see, I am not a stranger to haute cuisine. I just keep it in its place and indulge in that kind of meal when the time is right or there is an occasion to celebrate.

Movement and physical activity and the gym

Let's now talk about physical activity. I do not regard movement as having anything much to do with weight loss. As I have said elsewhere, eating and weight loss must be kept completely separate from the reasons to move. I do physical activity to allow me to perform all the daily functions that I enjoy. In the winter I go to the gym about every fourth day. I hate gyms. They all seem to adhere to the lowest common denominator for public gathering places. They are geared to the young. There is no place here for any kind of aesthetics. Nearly everything is black. Music blares with a tempo that is almost always the wrong one for the exercise I am doing. At the gym where I work out, music comes on the cheap from commercial radio. Thus, in addition to inappropriate music there are irritating adverts every few minutes. This atmosphere is supposed to increase my ability to concentrate. I come to the gym to do my exercise as quickly as possible in order to escape. I do not come to be entertained.

In the gym

When in the gym, I cycle for about 40 minutes at a level that suits me. I then do tummy exercises which consist of various forms of sit ups. I like the plank. Next, I do 30 reps of weight-lifting exercises, working my chest, deltoids, rhomboids and long back muscles, biceps and triceps, in that order. I do not stop but go from one exercise to the other. I do everything at a fast pace. I am not interested in bodybuilding and I am very conscious of the fact that all physical activity regimes must include resistance exercise. I assure you that there is no substitute. Resistance exercises are the tried-and-tested method to maintain muscle strength as we pass through the decades.

I suppose I will use up about 300 to 400 calories while I am there, I do not keep close watch. In any case I will have to replace half the calories I have used pretty quickly because of the total lost, glucose makes up one half and fat the other half. We need glucose to maintain our blood sugar so that we do not become hypoglycaemic. I can replenish the used calories with a drink and a sandwich. I never ever tell myself to exercise more if my trousers get slightly tight. I adjust my food intake. That way I am in control. After all these years I still use my waist to judge my weight.

In the summer the tempo of my activity changes. But I am finding it more difficult. There is a tendency when getting older, and when the slippage in ability appears, to work harder to rectify the loss. This is a bad choice on a grand scale. You cannot work against the inherent ageing process. You can only accommodate it. I cycle. My speed is well down from before, and the distances are shorter. It is annoying but does not really upset me. I am still moving. I love the feeling of wind in the face, moving at a human speed generated only by my muscles.

Round and about with June

June is two-and-a-half years older than me and is within touching distance of 90. She has her own routine for breakfast and lunch. However, she needs to adjust her intake to match her own energy expenditure. Obviously, it is different from mine. This difference again shows the difficulties of being prescriptive. If you demand specific exercise advice then you have not yet learned to factor in your gender, weight, age and physical activity. In addition, as one gets older, physiology changes more quickly. We do have dinner together. This is our time for planning, getting up to date and ensuring all is OK with the family.

June has her own favoured activity pattern. She swims with her friends and attends a dance/aerobic session once a week. Our interests intersect with walking. On our holidays, which we always take together, we spend much time hill-walking and exploring the countryside wherever we happen to end up. We will walk, not march, about 10–12 kilometres in a day. In our seventies we walked the South West Coastal Path in England, the Kepler Track in New Zealand, the mid-Drakensberg

Mountain paths in South Africa. In our mid-eighties, we are soon heading for the mid-Alpine pathways in Switzerland. Will we be successful? Who cares? Trying is the aim. How long will we keep trying? Till we drop.

Some advantages of keeping healthy at 80 and beyond

What do we gain from our efforts? Well, we are medication-free and live our independent life. June recently had a hip replacement. You cannot legislate for wear and tear. Because of her physiological status she is recovering rapidly. We were able to manage without outside help. The value of our independence is immeasurable. The physiotherapists who treated her were amazed at how well she performs tasks at the ripe old age of 86. They remarked that she behaves like a 60-year-old. This remark, as we have previously discussed, shows a lack of appreciation of the effects of the trinity on the ageing process.

CHAPTER FOURTEEN

A Final Look at Ageing Well and Wisely

This chapter intends to refresh your memory about everything you've just read. Please don't be offended by the suggestion that you might already need a reminder.

We have seen that there are three prime conditions that must be fulfilled in order to enjoy a healthy old age. I refer you back to the WHO definition as to what I mean by a healthy old age. Remember, it involves body and mind. This book has been centred on the trinity of factors – exercise, food and the mind – necessary for healthy ageing.

How long have these ideas been around? Well, Plato said, 'The part can never be well unless the whole is well.' It is rather depressing to pass shelves of books in which *only* one of the three factors has been singled out and expanded upon. This leaves the impression that the other two are of lesser importance. I cannot help feeling that these books are written by authors who have not yet had to face the harsh realities of failing physical strength and inappropriate weight gain that still lie before them as they enter the later decades of their lives.

Physical activity is not a cure-all

Perhaps by now you think that I am convinced that keeping to the trinity will protect you from all the vicissitudes of life? I certainly do not. I have, during this decade and some previous ones, seen many changes. Tragedies occur that are unpredictable and sometimes inexplicable. A few of my cycling colleagues are gone. There is also no accounting for wear and tear. Knees and hips can suffer from arthritis, especially if you become obsessed with exercise and overdo things on a daily basis. There is also no accounting for human biological variation. It is probable that not all people react to exercise in the same way. However, that should not be your focus. Compose your mind, set your goals and go for it.

I regard the effects of the trinity on health in the same way I think of the effects of not smoking. Giving up cigarettes saved many lives. But it did not totally remove the possibility of getting lung cancer. Even people who never smoked sometimes get the disease. Biology and disease are very complicated. So, adhering to the trinity is not a cast-iron guarantee of avoiding every misfortune, but without a doubt your chances of getting a disease are markedly decreased. That is the best anyone can hope for.

But, if you remain overweight and inactive, the odds are that you will spend approximately ten or more of your final years under medical treatment. How many visits to the doctor will this entail? How many drugs? How much support staff? How intensively will your poor family be involved? Apologies to John Donne, but no person is an island and the ripples from our independently made bad choices will spread and envelop our nearest and dearest. In contrast, my morbidity time is likely to be compressed. How do I know? Take a look at the figures for life expectancy at my age.

Colour Red: Exercise

The first colour mentioned is movement, physical activity, exercise. The reasons for activity are lodged deep within ourselves. It is a necessity driven by our evolution. It was not invented in this modern era. Everyday physical activity for the vast majority of us has little connection to athletic achievement. It is rooted in a basic need and the innate ability to chase and kill prey in order to live. Our bodies have no idea that we live in a new era in relation to how we obtain food. We must enjoy our chosen activity. If you like what you are doing, there is no stopping you. Your body will lift you to new heights.

Colour Blue: Food

The second colour is concerned with eating. When we overeat, we store the excess calories as evolution dictates – to sustain us during famine. Unfortunately, times have changed, and food is everywhere at all times. You cannot walk the high street without passing a source of food every 10 metres or so. In our times there is probably no need to store more than a day's requirement.

Colour White: The Mind

It is clear that implementing red and blue is not going to happen without mental fortitude. Not being mentally prepared means failure. That's it. If you get your mind together, you will secure the red and the blue, and you will feel better and be happier. The inter-relationships have a pleasing symmetry. Too little physical activity and there is trouble; too much eating and there are problems. Too little mental commitment and the other two will fade.

The Mind–body loop

Remember that physical activity, proper eating and a healthy mind require application over a lifetime. Stop and every gain will quickly fade away. When you embark on a healthy lifestyle, you must accept that a lifetime of dedication is necessary. The evidence for this is everywhere. In terms of integrative physiological function, there is no separation between mind and body.

Other factors

I realise there are many other elements that must also be in place. These are less easy to measure but good housing, comforting family relationships and stress-controlled work are all part of the picture. How do we ensure that these conditions are available? I have no idea. These are essential but well beyond the confines of this book.

We should never label healthy ageing as a disease

As we pass from one decade to another are we living less or dying more? You decide. Whatever you decide, it must be emphasised that we cannot and should not describe the process as a disease. There is no tablet, no operation, no medical intervention and no religious ethic that can halt the process. The strange idea that this inherited process is a disease is pervasive and wrong. This idea, in its wrongness, distorts society's approach to ageing.

If I was czar

Here goes . . . I would start by renaming the present NHS the National Disease Service, because that is what they do. I would set up a parallel system called the National Health Service. This organisation would be devoted to promoting healthy living. The twenty or so conditions wrongly called 'diseases of ageing', or perhaps better referred to as Exercise Deficiency Diseases, would be allocated to this new NHS. The funding of this organisation would be given priority over research devoted to the development of new non-curative pharmaceutics.

My pre-war cohorts and I are a lost generation. I would concentrate resources on ensuring that the present generation of 40-year-olds are taught not to ignore their own inevitable descent into old age. They are the problematic future not us present oldies.

The Lazarus Strategy

In particular, I would try to get the medical profession to understand the role of the trinity in preserving health. Indeed, I would make knowledge of the effects of physical activity, mind and diet an integral part of the medical curriculum. Gym personnel should become the first line of defence against an unhealthy lifestyle. However, they must be trained on the physiology of healthy ageing. I would insist on gym personnel keeping verifiable records on all their clients to ensure that physiological function is monitored and documented as their clients age.

Deep thought is required to change the ambiance of gyms. It would be a great help if gyms separated their bodybuilding facilities from resistance exercise areas designed to maintain function and health. Those aged who exercise for health do not require everything to be black and ageing people with small statures require equipment designed for their body shapes.

Preservation of health in ageing, rather than treatment of disease, should occupy centre stage. Thousands of lives can be prolonged by finding new and effective medications. However, the preservation of healthspan has the power to keep ageing people well and active right into their tenth decade. Millions of us can be potentially saved from long periods of morbidity. Name the drug that can top that.

I think I would also be inclined to cut down on research on longevity. This research is connected to the idea of removing death. The vast majority of life forms on our planet have evolved so that life and death are inextricably linked. Some molecular scientists and biologists have the ultra-optimistic view that it should be possible to unlink life and death. The problem is that there is no link. This figurative Corsican twin analogy is incorrect. Integrative physiology needs to take centre stage in research, education and application.

Saying farewell

Ordinary people are capable of extraordinary feats. They do not need special genes. Have another look in the mirror. You are looking at an ordinary human being who is capable of extraordinary things. Take a deep breath and make a promise to yourself: 'I am going to throw off the shackles that hold me down because I am overweight, I am going to develop my body to enable me to venture where I have never been. I am going to do that starting right now.' Join the 20 per cent of the population who control their ageing destiny to the limits that the ageing process allows.

We started this book with a quote from one Dylan (Bob), let us end with a quote from another – Dylan Thomas: 'Do not go gentle into that good night.'

CHAPTER FIFTEEN
My 'Ageing' Story

It is my belief that people over a certain age have acquired wisdom and a justifiable scepticism which make them resistant even to well-intentioned advice, especially from authors they have never met. And so, I have taken great care not to dictate behaviour, based on the belief that to do so would be insulting and, even worse, useless.

However, I think it could be helpful to tell you how I have lived at various stages of my life, assuming that you have not yet attained my years, in the hope that my experiences might be enlightening but not prescriptive.

Norman at 55: Tackling obesity

It is very difficult to uncover root causes of behaviour. Mostly the end product is the result of a steady accumulation of familiar, easily accepted habits. I come from a family that tended to be overweight, especially on my mother's side. That side of the family was described as being 'big-boned', whatever that was supposed to mean. There are umpteen euphemisms for fat. My brother was lucky, he had my father's genes so sailed through most of his life not having the same uneasy relationship with food that I had. My weight yo-yoed, but within limits that seemed acceptable to me. The major change in lifestyle came at about my 50th birthday and this change arrived at the same time as the children leaving home. Suddenly June and I redis-covered habits that had been parked for about 20 years. We liked good food and, with our new-found freedom, embarked on a food journey through the starred restaurants listed in *The Good Food Guide*. We had a lovely time. Other eating habits also sneaked in: fruit cake with our tea, malt loaf after dinner and chocolates around the house.

Creeping obesity

Despite this, we remained fairly active. Our main physical activity entailed walking in the UK, on the Austrian Hills and Italian Alps. Our UK walks are about 10 kilometres a day. We selected some of the most attractive long-distance pathways and are determined to finish those we have chosen. We did not walk these paths in one go. We watched the weather and when the outlook was good; off we'd go for about three or four days of walking. We started walking the South West Coastal Path. All accompanied by lots of good eating. Despite this activity I was putting on weight. A contributory factor could be that when I was bored, I had begun to use eating as a way of pass-ing the time. In retrospect, eating was central to my lifestyle so how else was I going to fill the time but by reverting to ingrained habits? It is disconcerting to think that my expand-ing waistline did not impinge on my self-image. I remember one occasion going into a men's outfitter in Regent Street in London in order to buy a new pair of trousers. I confidently asked the person serving for a pair with a 34-inch waist. His eyebrows lifted but he did what I requested. They were much too small. My waist measurement had always been 32–34 inches, clearly there must have been a change in the cut since the previous time I went shopping for trousers. Another pair, this time 35 inches, was produced. My chagrin was complete when they too did not fit. The next size up did fit, but unable or perhaps not wanting to process the information I was getting, I exited the store. There are many ways of wearing trousers that do not fit. A well-favoured option is to wear a belt and allow the fat to droop over the belt. One can then confidently claim that one's waist measurement had not changed for years.

The dawning of self-awareness

We were on a walking holiday in Switzerland, a holiday not unlike the many other breaks we have taken. Peeking figuratively around the corner I could see my 55th birthday coming like an express train. It had been a bright sunny day and June and I had spent the whole day on the heights chasing butterflies and Alpine flowers. We came down towards sunset, that great time of day when the tempo of life seems to slow, especially in the mountains. I had a shower, changed and went into dinner. I ordered soup for my starter. I cannot for the life of me remember the rest of the meal. Why does soup remain fixed in my mind? The soup arrived and I unravelled the napkin and looked down to place it on my lap. Where my thighs should have been lay a cloud of lard. I realised that I had become that ubiquitous caricature of a 50-year-old male that I had always been disparaging about. The man who appears to have swallowed a gigantic, indigestible pear. Why this day, why this time, I have no idea. Somewhere in the interstices of my brain my self-awareness of my body shape must have begun to begin to assert itself. I can only guess. I refolded the napkin and laid down the spoon. These two mundane actions represented to me the symbolic close of an era. I would never be fat again.

Losing weight

How should I go about losing my extraneous blubber? If you visit a bookshop or look on the internet you will find those hundreds of books with each author putting forward their weight-loss method as being great. How can this be? Does each author know a secret of nutritional biology that is opaque to the rest? I cannot believe that. On distilling the information, it appeared to me that weight loss depends upon eating less energy than you use.

Self-reliance and the will to do what I want to do are my foundation. After a bit of cogitation and discussions with June I opt for a radical approach. I want to lose weight quickly, that is my main drive. I also know I will not be able to do this on a balanced diet but I am sufficiently aware of the side effects. My tummy informs me I have at least a month's energy stored in fat. I am going to set a target. I reckon I need to diet for about ten weeks in order to lose well over 10 kilograms. This works out slightly more than a kilogram per week. I believe I can handle that. Obviously, this radical approach is not for everybody and I do not advise it unless you know what you are letting yourself in for. If in doubt, ask somebody who is knowledgeable and preferably someone who has been successful in losing weight. The trick, if there is one, is not to blindly follow what I or anyone else did. I am going to chart my own course and use the innate intelligence that I have been given. I am going to adopt a regime that suits me.

To help me along I keep a diary in which I note down everything I eat. I'm going to eat between 700 and 800 calories a day. To ensure I do not break down my muscles for energy I am also going to eat 150 grams of protein a day. I eat fruit, cottage cheese, green vegetables, tomatoes and cucumber. I continually monitor the number of calories I am eating. I do not weigh myself. My target is my 32-inch trousers that lie neglected in a dark corner of my wardrobe. I feel hungry but I reckon feeling hungry is part and parcel of the whole process. I do notice as the weeks pass that I am better able to handle the feeling. In any case I am well aware that feeling hungry while my tummy still bulges is a false signal. I know no harm will come if I ignore it. Three times a week, to ensure I am preparing myself for a slimmer body, I attend the gym and cycle for about 20 minutes. I do not regard the cycling as part of the weight-loss programme but as forward planning for better things. The fat disappears. I flaunt my 32-inch trousers.

At normal weight

I now begin to adjust my food and my intake. I increase my calorie intake to somewhere between 1,600 and 1,800 calories. I think that is about right for me. I also slowly introduce a more balanced diet. I do not let my vigilance wander. I still monitor my calorie intake very closely. My regime has allowed me to develop a very good understanding of the different calorie and nutritional values of many foods and I reckon in about another month I will be able to judge portion sizes and their calories without much trouble. Here I am at 55 years of age having finally got my eating under control. I now need to choose a physical activity that suits my particular nature and inclination. After trying a number of options I finally rediscover the bicycle.

Norman at 66: Discovery of the dragon

Holidays

June and I travel extensively and have no problems with reaching our destinations. We carry what we need in small rucksacks. This is to avoid having to check baggage and wait endlessly at the carousel. We are off the plane and out of the airport in the shortest possible time to continue onward by bus or train to our destination.

Walking is our pleasure; however, we do not go for concentrated hikes. Instead, we spend time looking around and enjoying our environment. We are busy ticking off various long-distance pathways but without the frenetic attitude that seems to accompany many long-distance walkers. June introduced me to walking and I really like this activity. One moves through the landscape at a human pace with the result that there is interaction with the surroundings, rather than a passing glance as happens in a car. If flowers, butterflies and birds are added to the mix then every walk becomes an adventure of discovery. We walked the Kepler Track in New Zealand where all walkers need to be self-sufficient in food. As I have already described, we also travelled many long-distance paths in the UK. Long may this continue.

My 'Ageing' Story

My body

I am of average height. I would expand that and say that most things about me are average. If you passed me in the street you would pay as much attention to me as you would to the street sign. Maybe less. When I was young and played games with playmates, I would not suffer the humiliation of not being picked for a team, but my selection was somewhere in the middle ranges. That was OK by me.

As I matured, I tried various activities. No matter what game I tried I just did not seem to have the correct set of limbs to play as I pictured myself playing the game. I stumbled into skiing, not exactly a mainline sport in the UK, and found to my surprise that I was capable of easily orienting my body in order to keep my centre of gravity where it needed to be despite the multitude of twists and turns. But skiing for a worker is only viable for about two weeks of the year, and in order to ski I needed to travel to high mountains outside of the UK. Not a good logistical choice but a great sport despite the obstacles.

The main problem is that two weeks of frenetic activity a year is not really a recipe for long term fitness. I tried squash and there again I was average. I could never consistently hit that sweet spot on the racquet that makes the ball travel like a rocket. This thud produced by the ball on the racquet is very similar to the noise that occurs when a golf club hits the ball just right. Anyone who has played will know what I mean. My trouble was that I could not work out how to do this on a regular basis. The overall result of all these average performances was that I kept reasonably fit without developing a passion for any of my sporting dalliances.

The bicycle

There I was, aged 55, rummaging around in the store room when I came across a cycle we had purchased for my son in order for him to ride to the station to catch the train to school. As far as I remember he hardly ever used it. I would guess it was not a cool thing for a teenager to do. The last time I had been on a bike I was ten years old, living on the flatland of the Orange Free State in South Africa. Looking at this rather dilapidated bike I recalled only good times and decided I would see whether my childhood memories were correct.

In 2005, BBC Radio 4 listeners rated the cycle as their top technical innovation. You can gather why. It is non-polluting, long-lived, requires no fossil fuel, makes no noise and keeps you fit. It can be used as a commuter vehicle, will transport you to any place you desire, and you can talk to your mate while riding. Best of all, you can move through the landscape at a rapid pace but still fully engage with your surroundings. What is there not to like?

On the cycle

Like many happenings in life, my affair with the cycle kicked off with a minor event: I cycled round the block. Of course, I overestimated my pedalling ability and ended up back home about 20 minutes later completely exhausted. However, the speed with which the ground was covered, the wind in my hair, and the feeling of freedom caught my imagination.

In addition, for the first time in a very long time I felt comfortable in what I was doing. My body seemed to meld into the cycle. Gradually I began to ride longer distances.

I started out joining short jaunts organised by charities. These rides had navigational aids and car traffic was closely monitored. For a neophyte cyclist these rides were great for learning road craft.

I progressed to doing cycle touring on my own. I found I enjoyed my own company and that during the ride my mind went into free wheel, screening scenes of my life like newsreels in a movie. I crossed England west to east at about 100 kilometres a day. Then around Ireland at 120 kilometres a day, followed by the 1,200-kilometre ride from Land's End to John o' Groats at 200 kilometres a day. Though I was not built for racing, I was suited to riding at a steady speed over long distances.

Discovering the dragon

My cycle is forged from steel and I am sure that in every forged cycle there dwells a dragon. Different cycles have different dragons. Velodrome bikes are inhabited by dragons that awake, roar and subside. Classic cycle events have cycles populated by dragons that roar and breathe fire trying to burn their competitors.

Then there is my dragon. I discovered its existence quite by chance. I joined Audax. Audax, which means brave in Latin, is the name of the UK long distance cycling club. All round the kingdom, the club puts on events covering distances ranging over 100, 200, 300, 400, 600, 1,000, 1,200 and 1,400 kilometres. The great thing about Audax is there is no competition within events. All riders are given a minimum and a maximum time to finish. Minimum times are set at 15 kilometres an hour. This speed is within range of nearly every fit cyclist.

The Lazarus Strategy

One result is that Audax, for years, was one of the few sporting organisations in which men and women could ride together in the same event. The distance is the challenge and any rider finishing in the allotted time is given the same recognition as any other finisher. There is no extra kudos for coming in first, other than personal satisfaction and going home earlier. Here, again, my cycling falls into the category of being average. I cannot cycle sufficiently fast to be a racer, I am not an excellent hill climber and I have no real desire to be first out of the pack. But what this average body can do is endure. My physiology is built for endurance and I had found my place in sporting disciplines.

About a year into my Audax riding I entered a 300-kilometre event starting at 2 a.m. in the morning. It is a horrendous time. The time itself is not the most disconcerting aspect. It is what is going on under the skin and in the mind where the real challenges lie. Indeed, it is for these very reasons that event organisers addicted to masochism choose this time. Body temperature is at its lowest reading in the circadian rhythm that controls our physiological processes over 24-hour cycles. Hormones like cortisol are also at their nadir. The mind is sluggish, as are the muscles.

Prior to the start I set my cycle up, checking that I had my emergency rations, that my lights were working, that the tyres were at the right pressure, and I mounted the saddle. Sitting there, quietly waiting for things to begin, a remarkable feeling overcame me. I felt at one with the cycle. After years together we appeared to have come to understand and know each another. The saddle felt like home. With unquestioned certainty I knew that I had sufficient power in my legs to tackle all obstacles that lay before me. I had the extraordinary feeling that if I set off immediately I could ride forever. I sensed my dragon resting easily in that awkward trapezoid around which a cycle is constructed. Only a small dragon to be sure but a dragon nevertheless. Not showy. Resting there quietly, exhaling little plumes of smoke, as it prepared itself to go on forever if necessary, at a constant distance eating pedal cadence.

That dragon and I became the UK Audax Veteran Champion for cycling the most kilometres in a year as well as champion of my club and county. We rode 1,200- and 1,400-kilometre events together. If we had the time, we would have ridden the world. All this accomplished by an average body performing at average speeds, on an ordinary cycle inhabited by an ordinary dragon.

I have now been cycling for about 11 years.

Some logistics of Audax riding

Let me give you some insight into riding 600 kilometres. This distance in the UK is equal to cycling from Cambridge to York and back. At a minimum speed of 15 kilometres an hour you can quickly work out that riders are given 40 hours to complete the event. In order to ensure that all entrants are keeping to the designated route, controls are set up every 100 kilometres. This requires all riders to check in at the first control in about six and half hours (100 kilometers divided by 15 kilometre/hour). Every subsequent control also closes six and half hours after the last, so there can be very little hanging about. If you arrive later than the designated time the control will be closed and you go home.

I ride at 20 kilometres an hour, so I can accumulate about one and a half hours per 100 kilometres (100 divided by 20 = 5 hours). This accumulated time becomes very useful if I need a nap on the ride. Besides the distance, there are other totally uncontrollable variables. These include wind direction, speed, rain, ambient temperature and trying to navigate the route through the night without getting hopelessly lost.

In addition, each ride has its own organiser and, in general, organisers seem to come from a band of controlled psychopaths. Main roads are avoided except maybe in those early, weary hours before dawn. In general, the rides are either difficult or very difficult. The most difficult 600 kilometer ride for me was a ride beginning at Poole on the south coast. According to the organiser this ride had around 8,000 metres of climb spread over the 600 kilometres. Now that is what I call difficult. When faced with such challenges, every person needs a dragon. Where do you find your dragon? Find yourself, know yourself, and your dragon will appear.

Norman at 70: On a cycle ride

I'm in the Blackdown Hills deep in Devon. I started at 10pm at night at the sea, heading north. The gut-wrenching hills of the coast are far behind me. It is around 1am and the rain has been pelting down for hours. Just ahead are the hills. I am about to enter the dark, narrow, winding lanes that snake through the landscape. I am navigating by means of a cycle lamp and a head torch. Because of the high hedgerows I can see only about 10 metres ahead.

My 'Ageing' Story

The rain does not help. Rain in Britain bears no relation to the quality of mercy. It is cold and its icy drops have penetrated deep into my clothes. Cold air is heavier than warm and it flows off the hills to the valley floors where it lies in wait for night riders. I struggle up an incline and plunge down the other side into a blanket of cold air. The temperature drops about ten degrees. My fingers become like cold bananas.

The lane heads sharply up. Because of banana fingers I miss the gear change and come to a standstill. There is only one direction to go when a cycle suddenly stops on a hill. The rider falls over. I fall into a puddle. When righting the cycle, I see that the fall has bent the gear cog. The proverbial straw has appeared. In a rage I throw the cycle onto the lane verge, sit down in the rain and vow I will never cycle again. Taunton on the motorway, where there is a service station and eatery, awaits but is still 12 kilometres distant. Do I walk? Do I leave the cycle to save time? Do I just sit and cry? What else is there to do on a Devon lane past midnight? The lanes are not exactly highways of traffic; that is the reason I chose the route in the first place.

Over the hill behind I see a flash of car headlights. A car at this time of night in these hills! I position myself in the middle of the lane. The car approaches. It is a taxi. The lady driver is returning from delivering members of a hen party who had cleverly hired a taxi before they drank themselves into a stupor. Of course, she would take me to Taunton provided I had the fare. Charity does not begin in a cab. I hand over the money, dump the cycle in the trunk, get in and ask her to turn the heater to maximum.

I start to thaw and by the time I am dropped off at the eatery I am feeling human again. The damage to the gear is not terminal and I quickly put it right. I enter the eatery and have a meal. The food is awful but I do not care. Now is not the time to order a yolk-free omelette or to enquire whether the food is organic or whether the provenance of the sausage is known or whether the chips come from local potato farmers. As I scoff away, I begin to feel much better. So, I have had a little adventure, so what? The night was full of fury and sound, like bagpipes at war. But nothing actually happened. I am in one piece. I have a full tummy and the bike is rideable. I'll doze here in the warmth slowly drying.

When the first rays of dawn begin to drive the shadows away, I too will be chasing them into the new day. There are other challenges waiting on the horizon.

Norman at 72: How am I doing?

Intimations of mortality

In the past few months something has changed in my mind. Whereas previously the difficulty of the route did not even enter my mind I now am having self-doubts. I am beginning to question whether I really want to do a difficult ride. These self-doubts are destructive and are the first real signs of age creeping into the agenda. I sense the dragon is not happy. Where, it wants to know, is that first fine careless rapture that used to accompany every ride? The dragon was rough-hewn over many months out of grey cells somewhere in my brain. As my confidence grew so did the dragon until one day, with a final crack of the chisel, it emerged fully grown and fully functional. It is clear that ageing is not only a physical deterioration but there is also a strong mental component in the mix. I have grown weary and so has the dragon.

I mow the lawn every week, about a hectare in area. My left knee gives a twinge now and again. I do not think there is serious trouble there. When you have exercised for a number of years you can begin to differentiate pain produced by a serious condition from pain that is only a background niggle. I think this is a background niggle. I get them quite frequently now. Sometimes a muscle will spasm for no apparent reason.

I definitely cannot see as well as I did about five years ago, and I notice that I also cannot hear high-pitched notes. I have become short-tempered with the news. I cannot bear the ignorance of the presenters, especially when they are discussing a subject I know quite a bit about. I cannot absorb information with the speed that I am used to. I cannot balance on one leg with my eyes closed, and my normal speed of walking appears to have slowed, because younger people stride past me without seeming to exert themselves. I skip pages in novels. I cannot stand hearing about youthful or middle-age angst. I have been through all that. Why should I read about it now? I am trying to eliminate all those things I feel I must do out of duty. I do not wish to be bothered with situations which do not give me a modicum of pleasure. I am changing my wardrobe. I feel I need more colour in what I wear.

Norman at 84: Going downhill but still active

My food intake is still balanced and sufficient to my needs. For the past 30 years I have been cycling and now, sad to say, with less enthusiasm than before. My ability to climb hills is now really bad. I use those little cogs and the large rear ones. I am ashamed to admit that I do not ride in inclement weather. I wait for the sun. I know it is a bit of a cop-out but so what? I also pick my rides carefully. I live on the Surrey-Kent border and the terrain has a multitude of hills. I try to convince myself I should ride my age in distance, 84 kilometres. I realise this is foolish because, as my age goes up, so will the distance. With the decline in muscle mass has also come a decrease in the speed I can generate. When I first became aware of this I put it down to insufficient training. A classic mistake made by oldies. Trying to regain speed by increasing training was really stupid. It just made things worse. I was overworking muscles that were attuned to my age. There is no way to make age-adjusted muscles better. My VO_2 Max is now lower.

More changes

I do not ride in the winter. I cannot face the cold. In addition, my fingers are sensitive to low temperatures and they become immobile and painful. During the dark, cold days, I spend time in the gym. In the gym, I do some cycling just to keep the thigh muscles up to scratch but I also incorporate resistance training into my regime. I do repetitions rather than large weights, mainly because I cannot lift heavy weights and I definitely do not wish to create more muscle mass to carry around. Most of my weight-bearing exercises involve no more than 35 kilograms, while for my arms I try to curl no more than 10 kilograms per arm. I do as many repetitions as I can. The weights I use have decreased markedly over the years but especially since my 80th birthday. I accept there is nothing to be done. I will keep doing what I am doing for as long as I am able. These exercises maintain those muscles that keep us upright and prevent falls and injury. Becoming bent over is not a good position to be in. You are already on the way down. We tend not to notice these slow changes in posture that can take place as the muscles become weaker. If you lean forward your weight could pull you down and cause you to fall, especially when going down stairs. It is most important to keep those long back and tummy muscles operating efficiently as you age.

What is going on in my head?

How much exercise do I do? It varies as I grow older, but my guiding thought is to do as much as my fading body allows. I inevitably set goals for myself. Sometimes I know the goals are unrealistic, but who cares? The idea is to not hide behind your age. This is a battle and involves internal monologues with the bad and good cop. You must try to control those voices in your head that keep whispering how old you have become and how now is the right time to go home and lie down.

The ageing process at work

I keep adjusting my lifestyle to my diminishing physiology. I do not dig in the garden anymore. It hurts my back. I do not drive at night. I cannot cope with the glare of oncoming head-lights. I tire more easily and find that on some days I need an afternoon nap, something I promised myself I would never do. I absolutely abhor piped music which is all pervasive. I do not expect people to have my taste in music, and I would not foist my tastes on them unasked, but every 40-year-old who is in charge of some public place seems to believe that their taste in music represents the culmination of evolutionary artistic genius. I am forced to raise my voice to speak in a restaurant. I am forced to listen to adverts in the gym and to exercise to music, the beat of which is totally out of sync with my workout speed. Who thinks up this stuff?

Trouble on the horizon

Ageing in humans is a mind–body interaction. In order to keep my mind active, I still have a place at the research table, trying to make a useful contribution. I hope that when I feel that I am not doing my share, or that I am talking rubbish, I will pack up. I sometimes wonder whether age is directly affecting our ability to properly assess when we should stop. The House of Lords springs to mind.

June and I go to concerts and the theatre, and take a holi-day abroad every now and then. Flying to our destinations may be coming to a close but we have decided that taking the train might be the answer provided we are able to carry the luggage we require. We want to keep our independence. We choose places where there is easy walking and things to do. I will continue to exercise in ever-decreasing amounts. I hope I die walking or cycling. I definitely will not welcome death but I do not fear it. The rhythms of life on earth require it.

My mind and some thoughts

As I age it is becoming clearer just how crucial brain function is for healthy ageing. I list mental function as being one of the three colours needed for healthy ageing. Lately there has been a continual internal dialogue taking place inside my brain. This involves the timidity of age partly brought on by the apprehension that a shrinking physical ability engenders, and the intellectual frontal cortex trying to allay these fears with logic. Does anybody under 75 understand this problem? In non-exercisers both the brain and physical systems are at a nadir and if the logic part of the argument fades then all that is left is the timidity. I suspect that this fracture between body and mind function could be a factor in making old people fair game for the con artists or anybody else who wishes to exploit their vulnerability, including some charities who should know better. If only I had the power to enthuse people to fly the flag of three colours for healthy living.

Incidentally, when it can be shown that animals chosen as models for healthy human ageing can have this dialogue and in addition worry about the cost of a university education for their grandchildren then I might believe they are adequate models for human ageing. I am sure there is a molecular biologist somewhere that believes fruit flies are the species of choice to investigate healthy human ageing. A fruit fly that lives a full and fruitful (oops) life will surely be like a healthy human oldie. There is a universe of difference in using an animal model for studying a disease and using that same model to provide information on a function specific to a species, namely the no-disease process of healthy ageing in humans.

At work

The science prospers and seems to have developed a life of its own. The challenge of research excites me. The new, smaller dragon awakened by my scientific work puffs smoke and gently roars in my brain. Is there still time for it to become sufficiently mature to suffuse its warmth throughout my body? We shall wait and see.

CHAPTER SIXTEEN
Some Real-life Stories

Harry

I was a successful businessman but in retrospect I retired much too soon. I was not sufficiently clued in to realise that when a large chunk of one's time is removed it is very important to have something to replace what you have discarded.

I started smoking while still a young man in an age when smoking was at its height. Every movie had a smoking scene. Cigarettes were handed out like candy. I took up smoking because that was the thing to do. My mates were drinkers so I went along with that as well. I ate whatever was put in front of me, provided the vegetable portion was small. I could eat a T-bone steak of about 800 grams at a sitting. I did very little exercise but I did try to keep my weight under control.

In my late fifties I gave up smoking and excessive drinking. I kept to an evening beer. In my sixties knowledge about exercise became more widespread. I devised a four-kilometre walk which I did daily, accompanied by my friend. I thought the exercise might offset the various insults I had thrown at my body. The walk had a small incline in it which, when I was reasonably fit, I barely acknowledged. Sometime in my late seventies I began to get slightly breathless when I walked the slope. I visited my GP who referred me on to a specialist. I had narrowed coronary arteries and I needed a stent put in. This was done but after recovery from the operation my breathlessness increased. Again, a visit to a consultant. He informed me that I had the early signs of emphysema. It is too late to acknowledge what a terrible habit smoking is and how persistent the ravages it brings.

The emphysema has progressed courtesy of all those cigarettes. I catch colds very easily and inevitably an infection that needs treatment results.

At 80 I am now on oxygen full time. I am restricted and a prisoner of the equipment. My musculature has wasted away from the chronic anoxia. I think that I will end this short history here.

Fay

I am a happily married woman with children and grandchildren. I retired 20 years ago and recall looking forward to having more time to follow my interests. The natural world is very important to me. My early years had been spent in a rural world of hills and sea and I planned spending more time outdoors, more travel, and more involvement with local affairs, more time for visits to exhibitions, concerts, and to go to plays, opera and ballet.

Energy

All these plans needed energy! I consider myself reasonably energetic for I naturally move and walk fast. I am not a sporty type, meaning into tennis or running marathons. I keep limber by going to a weekly hour long 'keep fit' class in the village hall – the same class for over 30 years. As I get older, I try hard not to miss a week – my body seems to need it. Many of us have been going year on year and we have become friends, ready for a chat as well as the physical activity. We support each other. I think the regime is a good workout for all the muscles and joints of the body. It involves more stretching than vigorous cardiovascular exercise, all to music with a beat and ends up with a jazzy dance routine. I've added a weekly swim for the last ten years. I go with a group of older women who on the whole are strong swimmers. I have never had much aptitude for swimming, and now would be classed as a very weak swimmer, but it is an opportunity to get into warm water and move different muscles. I look forward to meeting friends. After a swim and hot shower, it's a good start to the week.

I feel fit and find it hard to believe I am 86. I have to admit that my body – particularly my back – is stiffer than ten years ago. I realise that a weekly workout or so is OK but not enough. I need a daily routine. For example, I usually do a few leg stretching exercises in the morning on the bed when my muscles are relaxed before I get up. It doesn't take long – maybe 5 or 10 minutes. While getting dressed, I add a few more stretches – arm swinging, bending my knees – anything to get the blood circulating. This is my way of saying hello to my body each day. I'll stretch at other times of the day, say, after lunch, watching television or in the bath, maybe mobilising my arthritic hands.

There is a pattern to most days in that I try to ensure that I am active outdoors before lunch. I look forward to a walk either towards the hills around where we live or to a nearby wood to happily walk for an hour or so. There is always something new and extraordinary to look out for. Each season has its beauty, from watching the unfolding of young leaves in spring to the changing colours of autumn. If not walking, I'll be gardening. In winter I opt for a road walk as footpaths can be slippery.

After lunch I rest, as I have not got the energy I had five years ago. My 'rest' involves me lying on the floor in my study listening to the news on the radio for half an hour or so depending on my mood. At the end I'll include a few floor exercises – routine-Pilates type – and on getting up I attempt to balance on either foot as in a Yoga routine. I can't say that at my age I am very successful! All these times of 'movement' may read like a chore and a bore, but really they are not. They are stitched into each day.

Social

Keeping contact with family and friends is important, though with many friends and family living at a distance, this is usually by regular telephone contact. Over the years, many of my friends are suffering from ill health and some are housebound, and so contact and support of friends becomes even more important. With those friends living nearby, we now meet at lunchtime, rather than dinner. Now that I have less energy, I have to think more carefully about the timing of entertaining. Meals are simpler. Our friends agree, we don't have to impress each other with our culinary skills. We get together for friendship. We eat out more often.

All the family live at a distance and I have an excuse to travel by train. This is my favourite form of travelling long distances. At other times I'll use the train to visit a school or college friend and spend a couple of nights in different parts of the country. It is stimulating travelling by train. There is time to reflect, time to watch the countryside, the crops in the fields, the size of fields, field margins, the nature of woodland. Even a short train ride from home into different areas of countryside, where there is an opportunity for a walk before lunch, is attractive. Most holidays with my husband involve hiking. If the weather forecast is fair we will take pot luck and go somewhere in the UK for a couple of nights. I haven't the patience to put up with a stretch of poor weather. I have to admit that the distance I walk has decreased over the last few years – I haven't the energy – and the loss of puff prevents me from climbing very steep hills. Thank goodness I have a willing partner who will offer to carry my rucksack.

And back to me

I haven't spoken of food but this is important. I buy organic as much as possible and we eat a lot of colourful fresh fruit and vegetables. I start the day with muesli and fresh fruit. And no day would be complete without a large freshly prepared salad with fresh herbs. I don't snack. It is something I have never enjoyed; it is not a conscious deliberate act. My weight is pretty constant, around eight and a half stone with a BMI (Body Mass Index) of 20. I consider myself healthy for my age although I have no idea what is going on inside my body.

I have had arthritis for years and have a chronic kidney disease which at the moment is stable. Because of this I don't take Paracetamol and Panadol for my arthritic pains. We have masses of books in the house and I enjoy reading. But I have to be careful, particularly in winter. It is easy to sit for long periods and get stiff. I try not to sit for too long at any one time. I follow a 'sit-stand-sit-stand' routine.

Twenty years on from retirement I have gained huge pleasure in bringing aspects of the natural world to audiences through talks, walks and articles. I recognise that I have less energy and have to husband my time. I need more time to recover from visits, do fewer things in a day. For example, on a visit to London; I used to go to an exhibition during the day and a concert at night. No more! I am aware that I have to plan more. My motto is 'Look After Yourself!', coined by the Health Education Council 40 years ago. It was a fantastic idea at the time and has stayed with me all these years.

Monica

I am in my late eighties and married with children and grand-children, all happily living within an hour of each other. Actually, I am widowed but I don't use that word unless filling in a form since I could sound a bit gloomy or unhappy (and I make sure that I am neither). I regard it as a simple fact which I deal with. All of this is so normal but what I find interesting is, given a selection of situations, the different ways people deal with them. My way is not necessarily the right way but it is a positive way and suits my personality. It helps that I was born an optimist which influences my choices in life. It also affects the way I deal with any downsides which life throws at me.

I had always assumed that if we set our own standards and pathways, we could enjoy a happy and fulfilled life. But, in our late forties, my husband was left (overnight) totally disabled both mentally and physically. Even more sadly, he lived for another five years which affected us all in our different ways and, as the mother, I needed to hold everything, and everyone, together. The children were still at school; many elderly relatives lived nearby and relied on us for help. We also had family businesses to run. So, I had to think again about our priorities in life and how we could deal with them. I could have run away from the situation – looking around at many of my friends I could foresee that this would have been their way of coping. My big problem was – and this was entirely self-inflicted – I had to do everything I possibly could for my husband. I felt that was the very least I should do.

Fortunately my energy and, even more importantly, my patience never ran out so we were all able to enjoy a happy and contented five years. We always smiled, quietly laughed a lot, gently teased, and only spoke about good news. Worries, grievances and deaths in the family were discussed in the privacy of another room. Also, fortunately, I outlived my husband which was my greatest worry, and I learned much later, it was a worry shared by the children. So, that sums me up. Half of me is very ordinary and the other half less so.

Appearance

I have always felt myself blessed that I grew up when I did. Albeit a wartime child, personal standards were high. We were all more aware of behaviour, self-discipline and personal grooming, etc. Appearance was more highly valued. 'They' always say that we sum up a new person within the first few seconds. To be well groomed was not expensive so it was a standard which could be enjoyed by all. It saddens me to visit large cities (which I love) where almost everyone seems to be dressed for the beach or for a building site. Do people actually wear these clothes to work?

Since commuters look the same, presumably they do, how happy I am that, not only did I wear pretty clothes with gloves and shoes all carefully chosen, but that the same style of dressing and colour is still my choice. Lifestyle is also choice. This choice should be independent of age, as far as possible. Try to use age sparsely as an excuse for not partaking in something new. Life inevitably slows down so instead of going upstairs two at a time, just go up more slowly and more gracefully. No huff and no puff. Think: 'shoulders back; pull in your tummy; shoulders and tail bone down'. Off you go, and you will be delighted if it seems easier. No more walking for fear of falling over? Just practice standing on one leg more often to improve balance – and confidence. No need to make a fuss, standing at the sink or cooker is always a good time.

Keeping going

Did your mother send you to dancing class? Start again. There are a huge variety of exercise classes. Have you heard of Silver Swans for the over 55s? If there is not a class near you, just ask local teachers of small children to add a class for adults. If they are a trained dancing teacher there is very little extra needed. Amazingly, I find I can't jump anymore and my knees aren't what they were. But there are other bits that work fine so I concentrate on those and love the feeling and achievement of ballet and moving to music. Remembering the sequence of steps is excellent for the memory too.

Feeling a bit housebound? The library is always a good source of information on days out which is all good for the mind. While at the library, find a new subject to study – never stop learning. Even better, find out if there is a class nearby where we can meet more people. Not a joiner? Just take a deep breath and go – they will be very happy to see you. So often I hear people say, 'I nearly didn't go but I'm so glad I did.' Feeling short of like-minded friends? A few odd days out to, say, National Trust properties is a good alternative to going on holiday. Whatever your interest is will attract like-minded people to talk to. Given up driving at night? Socialise at lunch time instead. And, now that more cinemas are changing to several screens, films and events are being shown earlier in the day. We have the choice of local live screenings from the National Theatre and opera houses, etc., both here and abroad, and many societies that now cater for the ageing population: jazz to opera, needlework to picture framing, art appreciation. These endless activities leave little time for sitting in an armchair. The University of the Third Age caters for a huge selection of interests. How lucky we are to be so spoilt for choice. Lonely? Do something for someone else. Do you have a phone? Use it and give a friend the pleasure of receiving a cheerful call.

Help other people. I have always worked for charities. One, which lasted for more than 50 years, involved driving a 5-hour commute arriving home at about midnight. So I gave that up very recently and, not surprisingly, was quickly asked to take on another nearer to home. There are always those who need help which might not be physically demanding, but keeps the brain alert.

Making choices

Choice, or maybe motivation, is there for us all to achieve a healthy lifestyle. If we are living longer, why not aim to be healthy too. Why wait until we are ill when we can foresee that smoking or overeating will inevitably cause a problem. A problem which, in turn, prevents us from enjoying earlier activities. Does this sound as though I have had a much easier life than yours? Not necessarily. Active and with a lot of interests maybe, but that makes life enjoyable. At the same time I have coped, on my own, with major surgery for cancer and nursed both sets of parents. This applies to so many of us.

'Never complain and never explain' is an old saying. Getting up and doing something makes for a happier life even though it is changing as we get older. Just adapt. Keeping a flexible, open and enquiring mind will find a way of going around a brick wall and finding a different route to enjoying the day.

References

Chapter 1

Age UK Briefing: Health and Care of Older People in Britain 2019, July 2019.

Chapter 2

World Health Organization. Ageing and Health, 5 February 2018.

Morris J. N. and Heady J. A. 'Mortality in Relation to Physical Activity: A Preliminary Note on Experience in Middle Age.' *British Journal of Industrial Medicine* 1953 Oct 10(4): 245–254.

Duggal N. A., Pollock R. D., Lazarus N. R., Harridge S., Lord J. M. 'Major features of immunesenescence, including reduced thymic output, are ameliorated by high levels of physical activity in adulthood.' *Ageing Cell* 2018 Apr 17(2). https://doi.org/10.1111/acel.12750

Chapter 3

The Sunday Times Alternate Rich List 2018.

Chapter 4

Kingston A., Comas-Herrera A., Jagger C. 'Forecasting the care needs of the older population in England over the next 20 years: estimates from the Population Ageing and Care Simulation (PACSim) modelling study.' *The Lancet Public Health* September 1 2018 3(9): 2468–2667.

Kingston A., Robinson L., Booth H., Knapp M., Jagger C. 'Projections of multi-morbidity in the older population in England to 2035: estimates from the Population Ageing and Care Simulation (PACSim) model.' *Age and Ageing* May 1 2018 47(3): 374–380.

Pedersen B. K. and Saltin B. 'Evidence for prescribing exercise as therapy in chronic disease.' *Scan. J Medicine and Sports Sci*. 2006 Feb; 16, Suppl 1: 3–63.

Booth F. W., Roberts C. K., Laye M. J. 'Lack of exercise is a major cause of chronic diseases.' *Compre Physiol* 2012 Apr; 2(2): 1143–1211.

Pedersen B. K. and Saltin B. 2015. 'Exercise as medicine – evidence for prescribing exercise as therapy in 26 different chronic diseases.' *Scan. J Medicine and Sports Sci*. 2015 Dec; 25, Suppl 3: 1–72.

Lazarus N. R., Izquierdo M., Higginson I. J., Harridge S. D. R. 'Exercise Deficiency Diseases of Ageing: The Primacy of Exercise and Muscle Strengthening as First-Line

Therapeutic Agents to Combat Frailty.' *JAMDA*, 2018 Sep, 19(9): 741–743.

'Exercise Tips for Those with High Blood Pressure.' Web MD, April 2018.

BBC News. 1 February 2019. 'More over '75s should take statins, experts say.' www.bbc.co.uk/news/health-47058919

Chapter 5

Blair S. N., Kohl H. W. 3rd, Paffenbarger R. S. Jr, Clark D. G., Cooper K. H., Gibbons L. W. 'Physical fitness and all-cause mortality. A prospective study of healthy men and women.' *JAMA*. 1989 Nov 3; 262(17): 2395–401.

Gries K. J., Raue U., Perkins R. K., Lavin M. K., Overstreet B. S. and more. 'Cardiovascular and skeletal muscle health with lifelong exercise.' *J Appl Physiol* 2018 Nov 1; 125(5): 1636–1645.

Pollock R. D., Carter S., Velloso C. P., Duggal N. A., Lord J. M., Lazarus N. R. and Harridge S. D. 'An Investigation into the relationship between age and physiological function in highly active older adults.' *J of Physiol*. 2015 Feb 1; 593(3): 657–80.

Duggal N. A., Pollock R. D., Lazarus N. R., Harridge S., Lord J. M. 'Major features of immunesenescence, including reduced thymic output, are ameliorated by high levels of physical activity in adulthood.' *Ageing Cell* 2018 Apr 17(2). https://doi.org/10.1111/acel.12750

Chapter 6

Sessa W. C., Pritchard K., Seyedi N., Wang J., Hintze T. H. 'Chronic exercise in dogs increases coronary vascular nitric oxide production.' *Circulation Research*, 1994 Feb; 74(2): 349–353.

NHS Report: Cut down on your calories: Eat well. www.nhs.uk/live-well/eat-well/.

Food Standard Agency/Public Health England (2014) *McCance and Widdowson's The Composition of Foods: 7th Summary Edition*. London: Royal Society of Chemistry.

de Andrade L. P., Gobbi L. T., Coelho F. G., Christofoletti G., Costa J. L., Stella F. 'Benefits of multimodal exercise intervention for postural control and frontal cognitive functions in individuals with Alzheimer's disease: a controlled trial.' *Journal of the American Geriatrics Society*, 2013 Nov; 61(11): 1919–26.

Pollock R. D., Carter S., Velloso C. P., Duggal N. A., Lord J. M., Lazarus N. R. and Harridge S. D. 'An Investigation into the relationship between age and physiological function in highly active older adults.' *J of Physiol*. 2015 Feb 1; 593(3): 657–80.

Chapter 7

Statista: 'Fitness industry in the United Kingdom (UK) – Statistics & Facts.' Published by David Lange Sept 19, 2019.

Blackwell D., Clarke T. C. 'State Variation in Meeting the 2008 Federal Guidelines for Both Aerobic and

References

Muscle-strengthening Activities Through Leisure-time Physical Activity Among Adults aged 18–64.' *National Health Statistics Report*; 2018 Jun; (112): 1–22.

American College of Sports Medicine (2017) *ACSM's Guidelines for Exercise Testing and Prescription Tenth Edition*. New York: Lippincott Williams & Wilkins.

Healthcare Quality Improvement Partnership's National Hip Fracture Database (NHFD) Annual Report 2018.

Miller W. C., Koceja D. M., Hamilton E. J. 'A meta-analysis of the past 25 years of weight loss research using diet, exercise or diet plus exercise intervention.' *International Journal of Obesity Related Metabolic Disorders* 1997 Oct; 21(10): 941–7.

Thompson D., Karpe F., Lafontan M., Frayn K. 'Physical activity and exercise in the regulation of human adipose tissue physiology.' *Physiol Reviews*. 2012 Jan; 92(1): 157–91.

Kraemer W. J., Volek, J. S., Clark K. L., and more. 'Physiological adaptations to a weight-loss dietary regimen and exercise programs in women.' *J of Applied Physiology*. 1997 Jul; 83(1): 270–279.

Wu T., Gao X., Chen M., van Dam R. M. 'Long term effectiveness of diet-plus-exercise interventions vs. diet-only interventions for weight loss: a meta-analysis.' *Obesity Reviews*, 2009 May. 10(3): 313–323.

Martinez-Velilla N., Casas-Herrero A., Zambon-Ferraresi F., and more. 'Effect of Exercise Intervention on Functional Decline in Very Elderly Patients during Acute Hospitalization: A Randomized Clinical Trial.' *JAMA Intern Med*. 2019 Jan; 179(1): 28–36.

Isquierdo M., Rodriguez-Manas L., Sinclair A. J. 'What is New in Exercise Regimes for Frail Older People – How

Does the Erasmus Vivifrail Project Take Us Forward?' *Journal of Nutrition, Health and Ageing*, 2016 20(7): 736–737.

Chapter 8

Davey, R. C. 'The obesity epidemic: too much food for thought.' *British Journal of Sports Medicine*. 2004: 38(3): 360–363.

House of Commons Library Briefing papers SNO3336: Obesity Statistics. August 6 2019.

NHS digital: Statistics on Obesity, Physical Activity and Diet, England 2019.

Garnett T. Food Climate Research Network: 'What is a sustainable healthy diet? A discussion paper.' April 2014.

Alhassan S., Kim S., Bersamin A., King A. C., Gardner C. G. 'Dietary adherence and weight loss success among overweight women: results from the A to Z weight loss study.' *Int J Obesity (Lond)*, 2008 Jun; 32(6): 985–81).

Food Standard Agency/Public Health England (2014) *McCance and Widdowson's The Composition of Foods: 7th Summary Edition*. London: Royal Society of Chemistry.

Tsai A. G., Wadden T. A. 'The evolution of very-low-calorie diets: an update and meta-analysis.' *Obesity* 2006 Aug; 14(8): 1283–93.

Robineau D. 'Ageing Britain: two-fifths of NHS budget is spent on over-65s.' *Guardian*: 1st February 2016.

Chapter 9

Bhella P. S., Hastings J. L., Fujimoto N., Shibata S., Carrick-Ranson G., Palmer M. D., Boyd K. N., Adams-Huet B., Levine B. D. 'Impact of lifelong exercise "dose" on left ventricular compliance and distensibility.' *J Am Coll Cardiol* 2014 Sep; 64(12): 1257–1266.

Lazarus N. R., Harridge S. D. R. 'Declining performance of master athletes: silhouettes of the trajectory of healthy human ageing?' *J of Physiology*, 2017 May; 595(9): 2941–2948.

Chapter 10

Lazarus N. R., Lord J. M., Harridge S. D. R. 'The relationships and interactions between age, exercise and physiological function.' *J of Physiology*, 2019 Mar; 597(5): 1299–1309.

Chapter 11

Demetrius L. 'Of mice and men. When it comes to studying ageing and the means to slow it down, mice are not just small humans.' *EMBO Reports*, 2005 Jul; 6: 539–544.

Biological Effects of Dietary Restriction. ed. Fishbein L. Springer: 1991.

Ravussin E., Redman I. M., Rochon L. et al. 'A 2-Year Randomised Controlled Trial of Human Caloric Restriction: Feasibility and Effects of Predictions of Health

Span and Longevity.' *J Gerontol A Biol Sci Med Sci*. 2015 Sep; 70(9): 1097–1104.

Office for National Statistics. 'Life Expectancies: How long, on average, people can expect to live using estimates of the population and number of deaths.' 2018.

Cotman C. W., Engesser-Cesar C. 'Exercise enhances and protects brain function.' *Exerc Sport Sci Reviews*, 2002 Apr; 30(2): 75–79.

Chapter 12

Donato A. J., Tench K., Gluek D. H., Seals D. R. 'Declines in physiological functional capacity with age: a longitudinal study in peak swimming performance.' *J Applied Physiology*, 2003 Feb; 94(2): 764–769.

Baker A. A. and Tang Y. Q. 'Aging performance for masters records in athletics, swimming, rowing, cycling, triathlon and weightlifting.' *Exp Aging Res*. 2010: 36(4): 453–477.

Lazarus N. R., Harridge S. D. R. 'Declining performance of master athletes: silhouettes of the trajectory of healthy human ageing?' *J of Physiology*, 2017 May; 595(9): 2941–2948.

Acknowledgements

Thanks to Steve Harridge, friend and ace physiologist.

Thanks to Harry, Fay and Monica for their stories.

Thanks to my editor Bill Tonelli, who sought and brought the nowhere man to life.

Thanks to Euan Thorneycroft from A M Heath & Co, for finding the right trail.

To Lauren Whelan, Emma Knight and all at Yellow Kite, and to Kate Latham, for their support.

Index

Index

Index

and hospital admissions 86
incidence 86
and medical treatment 132–3
Norman's story 139–43
oxygen: and exercise 35–41
VO^2 Max levels 35–41, 101,
106–7

pain 25, 153, 164
Pavlov's dog 56, 90, 96, 97, 124
Pedersen and Saltin 24
Performance curve 118, 121
peripheral artery disease 8
physical activity *see* exercise
Plato 131
porridge 123–4
portion sizes 97, 143–4
prehistoric man 32–3, 45–6, 47–8,
77–8
processed food 96, 125
protein 97, 125, 143
pulmonary diseases 25

ready-meals 96, 125
relationships 134, 162–3
religion 56
repetitions, resistance exercises
72–3, 154–5
resistance exercises 71–3, 127,
154–5
restaurants 92, 97, 126, 140
resting heart rate 107
retirement 159
running 40, 65, 68, 107, 118

salads 91, 163
set point theory, exercise 101–3
shopping 53–4, 93, 97
sight 153
Silver Swans 166

skiing 145
small intestines 108
smoking 19–20, 132, 159–60
snacks 85, 89–90, 124
society, food and 56, 90
squash 145
stairs, running up 34–5
statins 29
stretches 161
strokes 8, 25
The Sunday Times 19
supermarkets 53–4, 93, 97
swimming 65, 67, 107, 116–18,
119–21, 161

television 85–6
temperature, body 148
Thomas, Dylan 137
trinity, healthy ageing 44–7
tuberculosis 57–8

University of the Third Age 167

vegetables 91, 97, 125, 163
vegetarianism 56
verbal fluency 106
Very Low Calorie Diet (VLCD)
95–6, 142–3
vision 153
VO^2 Max levels 35–41, 101, 106–7

waist measurements 92, 95, 127,
140–1, 143
Waksman, Selman 58
walking 49–50, 65, 66
goals 71
Harry's story 159–60
hill-walking 68, 128
long-distance walking 128–9,
140, 144